AF167650

Lecture Notes of the Institute
for Computer Sciences, Social Informatics
and Telecommunications Engineering   579

The LNICST series publishes ICST's conferences, symposia and workshops.
LNICST reports state-of-the-art results in areas related to the scope of the Institute.
The type of material published includes

- Proceedings (published in time for the respective event)
- Other edited monographs (such as project reports or invited volumes)

LNICST topics span the following areas:

- General Computer Science
- E-Economy
- E-Medicine
- Knowledge Management
- Multimedia
- Operations, Management and Policy
- Social Informatics
- Systems

Phan Cong Vinh · Nguyen Thanh Tung

Editors

Context-Aware Systems and Applications

12th EAI International Conference, ICCASA 2023
Ho Chi Minh City, Vietnam, October 26–27, 2023
Proceedings

 Springer

Editors
Phan Cong Vinh ⓘ
Nguyen Tat Thanh University
Ho Chi Minh City, Vietnam

Nguyen Thanh Tung
Vietnam National University - International
School
Hanoi, Vietnam

ISSN 1867-8211   ISSN 1867-822X (electronic)
Lecture Notes of the Institute for Computer Sciences, Social Informatics
and Telecommunications Engineering
ISBN 978-3-031-58877-8   ISBN 978-3-031-58878-5 (eBook)
https://doi.org/10.1007/978-3-031-58878-5

This Springer imprint is published by the registered company Springer Nature Switzerland AG
The registered company address is: Gewerbestrasse 11, 6330 Cham, Switzerland

If disposing of this product, please recycle the paper.

Preface

ICCASA 2023 (the 12th EAI International Conference on Context-Aware Systems and Applications), was held during October 26–27, 2023 at Nguyen Tat Thanh University in Ho Chi Minh City, Vietnam, in fully online style, due to the travel restrictions caused by the worldwide COVID-19 pandemic. The aim of the conference is to provide an internationally respected forum for scientific research in the technologies and applications of smart computing and communication. ICCASA provides an excellent opportunity for researchers to discuss modern approaches and techniques for smart computing systems and their applications. These proceedings of ICCASA 2023 are published by Springer in the series of Lecture Notes of the Institute for Computer Sciences, Social Informatics and Telecommunications Engineering (LNICST; indexed by DBLP, EI, Google Scholar, Scopus, and Thomson ISI).

For this twelfth edition of ICCASA, and repeating the success of the previous year, the Program Committee received submissions from authors in six countries and each paper was reviewed by at least three expert reviewers. We chose 14 papers after intensive discussions held among the Program Committee members. We really appreciate the excellent reviews and lively discussions of the Program Committee members and external reviewers in the review process. This year we had two prominent invited speakers, Issam Damaj from Cardiff Metropolitan University in the UK and Ngo Ha Quang Thinh from Ho Chi Minh City University of Technology, HCM-VNU in Vietnam. ICCASA 2023 was jointly organized by the European Alliance for Innovation (EAI) and Nguyen Tat Thanh University (NTTU). This conference could not have been organized without the strong support of the staff members of the two organizations. We would especially like to thank Imrich Chlamtac (University of Trento), Marica Scevlikova (EAI), and Stephen McGarry (EAI) for their great help in organizing the conference.

October 2023

Phan Cong Vinh
Nguyen Thanh Tung

Organization

Steering Committee

Imrich Chlamtac   University of Trento, Italy
Phan Cong Vinh   Nguyen Tat Thanh University, Vietnam

Organizing Committee

General Chair

Phan Cong Vinh   Nguyen Tat Thanh University, Vietnam

Technical Program Committee Chair

Nguyen Thanh Tung   International School, VNU, Vietnam

Web Chair

Nguyen Van Han   Nguyen Tat Thanh University, Vietnam

Publicity and Social Media Chair

Pham Van Dang   Nguyen Tat Thanh University, Vietnam

Workshops Chair

Hafiz Mahfooz Ul Haque   University of Central Punjab, Pakistan

Sponsorship and Exhibits Chair

Vu Tuan Anh   Industrial University of Ho Chi Minh City, Vietnam

Publications Chair

Phan Cong Vinh   Nguyen Tat Thanh University, Vietnam

Local Chair

Bach Long Giang   Nguyen Tat Thanh University, Vietnam

Technical Program Committee

Technical Program Committee Chair

Nguyen Thanh Tung   International School, VNU, Vietnam

Technical Program Committee Members

Anh Dinh	University of Saskatchewan, Canada
Ashish Khare	University of Allahabad, India
Bui Cong Giao	Saigon University, Vietnam
Cao Van Kien	Nguyen Tat Thanh University, Vietnam
Chernyi Sergei	Admiral Makarov State University of Maritime and Inland Shipping, Russia
Chien-Chih Yu	National Chengchi University, Taiwan
Dang Thanh Tin	Ho Chi Minh City University of Technology, Vietnam
David Sundaram	University of Auckland, New Zealand
Do Thi Thanh Dieu	Nguyen Tat Thanh University, Vietnam
Gabrielle Peko	University of Auckland, New Zealand
Giacomo Cabri	University of Modena and Reggio Emilia, Italy
Hafiz Mahfooz Ul Haque	University of Central Punjab, Pakistan
Harun Baraki	University of Kassel, Germany
Hiroshi Fujita	Gifu University, Japan
Huynh Xuan Hiep	Can Tho University, Vietnam
Hyungchul Yoon	Chungbuk National University, South Korea
Issam Damaj	Beirut Arab University, Lebanon
Kurt Geihs	University of Kassel, Germany
Le Hoang Thai	Ho Chi Minh City University of Science, Vietnam
Le Hong Anh	University of Mining and Geology, Vietnam
Le Xuan Truong	Ho Chi Minh City Open University, Vietnam
Manish Khare	Dhirubhai Ambani Institute of Information and Communication Technology, India

Muhammad Athar Javed Sethi	University of Engineering and Technology Peshawar, Pakistan
Ngo Ha Quang Thinh	Ho Chi Minh City University of Technology, Vietnam
Nguyen Thanh Hai	Can Tho University, Vietnam
Om Prakash	Hemvati Nandan Bahuguna Garhwal University, India
Pham Quoc Cuong	Ho Chi Minh City University of Technology, Vietnam
Rajiv Tewari	University of Allahabad, India
Shahzad Ashraf	Hohai University, China
Tran Huu Tam	University of Kassel, Germany
Truong Cong Doan	International School, VNU, Vietnam
Vu Tuan Anh	Industrial University of Ho Chi Minh City, Vietnam
Waralak V. Siricharoen	Silpakorn University, Thailand

Contents

Context-Aware Systems

Opinion Mining with Manifold Forests

Phuc Quang Tran[1] , Hanh My Thi Le[2] , and Hiep Xuan Huynh[3]([⊠])

[1] People's Police University, Ho Chi Minh City, Vietnam
tqphucth@gmail.com
[2] Faculty of Information Technology, The University of Danang - University of Science and
Technology, Danang, Vietnam
ltmhanh@dut.udn.vn
[3] College of Information and Communication Technology, Can Tho University, Cantho,
Vietnam
hxhiep@ctu.edu.vn

Abstract. Online reviews are becoming increasingly popular every day. They represent opinions and a wealth of information that can benefit organizations and individual consumers. However, studies on opinion mining have not focused much on classifying views according to the manifold to solve the problem of affinity between clusters of opinion, improving the accuracy, effectiveness, and generalizability of the modeling. In this paper, we have built an opinion mining framework with manifold forests to solve the influence of clusters of opinions based on the affinity between pairwise opinion points in each cluster and the relationship between different opinion clusters in large-scale data. In particular, we have focused on building a clustering trees ensemble and determining the affinity and distance of point pairs in feature space. Finally, the random forests are aggregated by ensemble methods such as stacking with a random forests classifier to identify opinion classification in reviews as either negative or positive. We used two datasets in the experiment to evaluate restaurants and hotels in two different scenarios, proving the effectiveness of the proposed model.

Keyword: Manifold forests · Clustering · Ensemble methods · Opinion mining

1 Introduction

With the explosive growth of global information [6], online applications on the internet are increasingly popular and effective for individuals and organizations. Increasing internet speeds, the appeal of social media sites, and e-commerce have resulted in a huge amount of informational data being reviewed online in the form of text. These reviews represent opinions [9] and a wealth of information that can benefit organizations and individual consumers. Among the techniques used in opinion mining, such as machine learning algorithms, lexicon-based approaches, and others, machine learning is commonly used in opinion mining. In machine learning algorithms, ensemble methods have many benefits to enhance the efficiency, performance, and generalizability of

P. Cong Vinh and N. Thanh Tung (Eds.): ICCASA 2023, LNICST 579, pp. 3–18, 2024.
https://doi.org/10.1007/978-3-031-58878-5_1

the model by dividing the original data into sub-datasets to match the basic learning models according to a certain method. The paper [10] presents an effective method for extracting opinion words to be used for opinion classification by ensemble learning [14]. Clustering algorithms are used to address the issue of processing unlabeled data. This paper [8] approaches the clustering method of product features for opinion mining. The proposed semi-supervised learning task uses the EM algorithm to solve the problem by improving labeled samples and allowing them to switch classes. The proposed method is evaluated effectively.

However, in the difficult case where opinion data comes from a combination of several non-linear dimensional manifolds, it is difficult to use conventional clustering algorithms for opinion clustering. Therefore, the use of manifold clustering [4] is a necessary solution to solve the problem of unsupervised machine learning in non-linear manifold space. In the article [3], the manifold clustering approach in deploying the manifold forest model has demonstrated that the manifold forest algorithm can close the gap with neural networks. In particular, the paper [1] has solved the general problem for clustering purposes by manifold clustering, which has superior results compared to worm subspace clustering based on autoencoders. The proposed Neural Manifold Clustering and Embedding (NMCE) method is closely related to and further understood with some self-supervised learning (SSL) methods. Currently, opinion mining studies have not focused on solving the problem of majority influence on the affinity between pairs of opinions in an opinion cluster or the affinity between different opinion clusters in large-scale opinion data.

In this paper, we propose an opinion mining framework with manifold forests to solve the problem of majority influence on the affinity between opinion points pairwise in an opinion cluster and the relationship between different opinion clusters in feature space. First, we focus on building a clustering tree ensemble. Second, we determine the affinity and distance of opinion point pairs in feature vector space. Third, we build independent random forests on the clusters. Finally, the random forests are synthesized by ensemble methods such as stacking with random forest classifiers to identify opinion classification in reviews as either negative or positive. We use two datasets of hotel and restaurant reviews to experiment with the model. The model's performance results are more accurate than the baselines, demonstrating the effectiveness of the proposed model. The framework of the proposed model is described in Fig. 1.

In addition to the content of Sect. 1, the paper is organized as follows: Sect. 2 discusses opinion modeling as the basis for opinion discovery in Sect. 3; Sect. 4 converts the results of opinion discovery into the opinion quintuples matrix; Sect. 5 presents the main content of opinion manifold forests; Sect. 6 evaluates the results of the model; Sect. 7 discusses opinion summarization; Sect. 8 experiments with the model on two different scenarios and discusses the results; Sect. 9 concludes the paper.

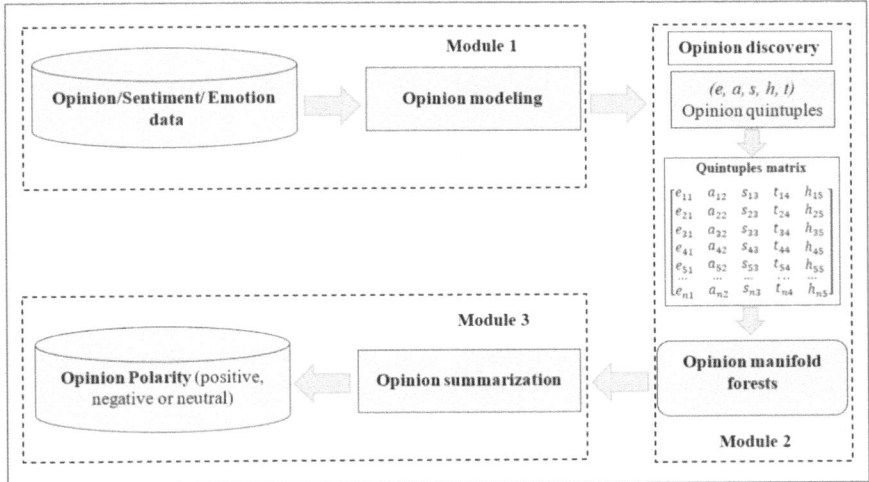

Fig. 1. The framework describing the proposed model

2   Opinion Modeling

2.1   Opinion

An opinion [9] modeled as a group of four components is called a quadruple (g, s, h, t), g is the opinion target, s is called the sentiment of the opinion towards the target g, and h is the person or organization giving opinion is called an opinion holder, t is the time to give the opinion of the person or organization. An opinion holder is an individual or organization that states an opinion. The time of opinion is the posting time or statement of the opinion holder.

Opinion Target. [9] The target of opinion or sentiment is the sentiment expressed to an entity or an aspect of the entity.

Sentiment of Opinion. [9] Sentiment of opinion is modeled as a triple quadruple (y, o, i). Sentiment types y can be classified into several categories based on linguistics, psychology, and classification based on consumer research. On the basis of consumer research is divided into two categories: rational emotions and emotional emotions. The sentiment of opinion is expressed through sentiment orientation o or opinion polarity can be negative, positive, or neutral. The sentiment of the opinion also depends on the intensity of the sentiment i, i.e. the different strengths and weaknesses of the sentiment. In practical application, it is possible to represent sentiment intensity as a number of discrete ratings based on two sentiment types such as a rating of 1 to 5 stars.

2.2   Simplify Opinion Definition

An opinion [9] can be defined simply and in more detail as a quintuple (e, a, s, h, t) where e is the entity to which the target is directed or the target is directed to the aspect a of the entity e, s is a sentimental orientation or polarity of opinion whose value is negative, positive or neutral, possibly also a 1 to 5-star rating. If sentiment reflects directly to the entity, the aspect represented is GENERAL. In this case e and a also express sentiment target.

In addition to the five components quintuple (e, a, s, h, t) of an opinion, there are reasons and qualifiers. A reason for an opinion [9] is the cause or explanation of the opinion. A qualifier [9] of an opinion limits or modifies the meaning of the opinion.

2.3 Opinion Entity

An opinion entity [9] e is represented by itself as a whole and a finite set of opinion aspects $A = \{a_1, a_2, \ldots, a_s\}$. Each aspect $a \in A$ of entity e can be expressed with any one of a finite set of its aspect expressions $\{ae_1, ae_2, \ldots, ae_n\}$.

2.4 Opinion Document

An opinion document [9] D includes a finite set of opinion entities $\{e_1, e_2, \ldots, e_r\}$ and a subset of opinion aspects $\{a_1, a_2, \ldots, a_s\}$ of each opinion entity. The opinion is derived from a finite set of the opinion holder $\{h_1, h_2, \ldots, h_p\}$ and at a particular time t.

3 Opinion Discovery

Given an opinion document D, opinion discovery on opinion document D is performed on the order of eight tasks as follows [9]:

First, extract the expressions of entities in document D and group similar entities and group entities into clusters or classifications. Each entity represents an entity clustering expression. Second, extract the aspect expressions in document D and group the aspects into clusters similar to the first task. Third, extract the holder's expression of each opinion from reviews or structured data and group them. Fourth, extract the posting times of each opinion and normalize the times by different formats. Fifth, classifies the perspective of the aspect by identifying the aspect or entity that has a positive, negative, or neutral opinion. Sixth, synthesize the above tasks to form all groups of opinions quintuple (e, a, s, h, t). Seventh, to extract the opinion reasons for each opinion and group the synonymous reason into a cluster. Each reason expression represents a group of opinion reasons. The final, extract the opinion qualifier expression for each opinion and do the same as the seventh task.

4 Opinion Quintubles Matrix

Given an opinion quintuple (e, a, s, h, t), where e, a, s, h, t are the features of the opinion whose values are respectively $(e_{n1}, a_{n2}, s_{n3}, h_{n4}, t_{n5})$, where n number of sample values. The value of the opinion feature can be a discrete value, a continuous value, or a mixed value. Convert opinion feature values to numeric values to get an matrix O : n x 5, with n number of sample values of opinion features represented as follows (Fig. 2).

The matrix $\mathcal{O}$ is the input data to train the opinion manifold forest model that we present below.

$$\begin{bmatrix} e_{11} & a_{12} & s_{13} & h_{14} & t_{15} \\ e_{21} & a_{22} & s_{23} & h_{24} & t_{25} \\ e_{31} & a_{32} & s_{33} & h_{34} & t_{35} \\ e_{41} & a_{42} & s_{43} & h_{44} & t_{45} \\ e_{51} & a_{52} & s_{53} & h_{54} & t_{55} \\ \cdots & \cdots & \cdots & \cdots & \cdots \\ e_{n1} & a_{n2} & s_{n3} & h_{n4} & t_{n5} \end{bmatrix}$$

Fig. 2. Matrix of Quintubles

5   Opinion Manifold Forests

An opinion manifold forest is an ensemble collection of clustering trees that simultaneously considers the relationship between points in the opinion feature space. Therefore, the opinion manifold forests model needs to estimate the affinity or distance between opinion points to be able to preserve the distance of those data points after mapping. In Fig. 3 proposed the manifold forests for opinion polarity/opinion classification.

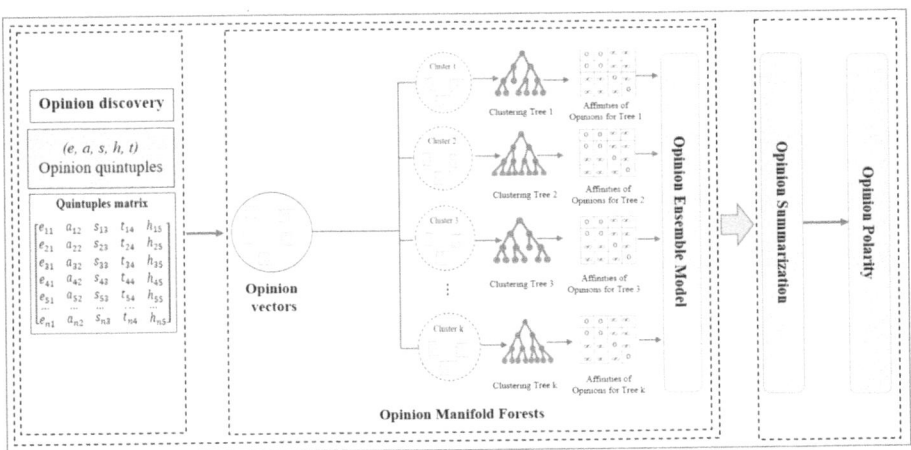

Fig. 3. The proposed model of the manifold forests for opinion polarity

5.1   Opinion Vectors

In the opinion feature space for opinion feature vectors $O = (e, a, s, h, t) \in \mathbb{R}^5$, each vector $o \in O$ is responsible for representing a point in the feature space and each vector opinion has five dimensions representing features for an opinion, each feature has a numeric data type and is one dimensional. These opinion feature vectors are used to train the opinion manifold forest model.

5.2 Opinion Manifold Forests Model

Given a set of k opinion samples $O = \{o_1, o_2, \ldots, o_k\}$ unlabled with $o_i \in \{R\}^d$, find a mapping $f : \{R\}^d \to \{R\}^{d'}, f(o_i) = o'i$ so that $d' \ll d$ and preserve affinity or distance related points in feature vector space. Each opinion input sample $o_i = (e, a, s, h, t) \in \{R\}^5$. The output is sentiment $s \in \{$positive, negative$\}$

The opinion manifold forest model is trained using the maximization of random nodes and trains the forest by maximizing the information gain measure. So, the process of splitting the jth node for optimization is done by the objective function as follows

$$\theta_j = \underset{\theta \in \tau_j}{\operatorname{argmax}} I(O_j, \theta) \tag{1}$$

with I is the continuous information gain as in the opinion density forest model [16]

$$I(O_j, \theta) = \log(|\Lambda(O_j)|) - \sum_{i \in \{L, R\}} \frac{|O_j^i|}{|O_j|} \log(|\Lambda(O_j^i)|) \tag{2}$$

where O_j is the set of opinion samples that split the jth node, O_j^i is the set of opinion samples on the path to the left and right of the jth node, $\Lambda(O_j^i)$ is the covariance matrix at node split jth.

Different from opinion density forests [16], the opinion manifold forest model estimates a measure of the affinity between opinion data points to maintain the distance between those data points after performing mapping using random forests to define opinion affinity. At the leaves of a clustering tree t defines a partition of the input opinion points.

$$l(o) : \{R\}^5 \to L \subset \{N\} \tag{3}$$

with l as the leaf node index and L as the set of all leaves in a tree. Each clustering tree t can compute the affinity matrix $W^t : k \times k$ with

$$W_{ij}^t = e^{-Q^t(o_i, o_j)} \tag{4}$$

where Q is the distance determined by the binary affinity of the pair (o_i, o_j) of opinions.

$$Q^t(o_i, o_j) = \begin{cases} 0 & if \ l(o_i) = l(o_j) \\ \infty & otherwise \end{cases} \tag{5}$$

where $d_{ij} = o_i - o_j$, and $\Lambda_{l(o_i)}$ is the covariance matrix linked to the leaf node by the point o_i.

Given a tree t and two o_i and o_j, if points o_i and o_j end up in the same cluster (leaf), then assign affinity equal to 1, distance equal to zero for the pair (o_i, o_j). Otherwise, assign affinity equal to zero, distance equal to ∞.

The opinion affinity matrix of the manifold forests is the ensemble affinity matrix of the trees of the entire forest of size T and is calculated by averaging [12] affinity matrix W^t from single opinion clustering trees

$$W = \frac{1}{T} \sum_{t=1}^{T} W^t \qquad (6)$$

6 Evaluation

The proposed model is evaluated through Table 1 called the opinion confusion matrix, which consists of four different combinations of opinion predictor value and actual opinion value. The classification performance of the model in this case is evaluated as either a negative or positive classifier [15]:

Table 1. Opinion confusion matrix

	Opinion predictor positive	Opinion predictor negative
Actual opinion positive	True Positive (TP)	False Positive (FP)
Actual opinion negative	False Negative (FN)	True Negative (TN)

Where True Positive (TP) represents the number of opinion samples with true positive opinion values that are predicted to be true positive. False Positives (FP) represent the number of opinion samples with true positive opinion values predicted that are not true positive. False Negative (FN) represents the number of opinion samples whose true negative opinion values are predicted to be neither true negative. True Negative (TN) represents the number of opinion samples with true negative opinion values that are predicted to be true negative.

Measures of accuracy, precision, recall, and F1 are used to evaluate the opinion classification performance of the proposed model.

$$Accuracy = \frac{TP + TN}{TP + TN + FP + FN} \qquad (7)$$

$$Precision = \frac{TP}{TP + FP} \qquad (8)$$

$$Recall = \frac{TP}{TP + FN} \qquad (9)$$

$$F1 = \frac{2 * (Recall * Precision)}{Recall + Precision} \qquad (10)$$

7 Opinion Summarization

Opinion summarization is an aggregation of user opinions expressed in reviews online. The major opinion summarization work often takes the entity-centric aspect of the entity as a hierarchy to create a specific summary consisting of documents related to an entity or an aspect of the entity. Such opinion summaries are called aspect-based opinion summaries or feature-based opinion summaries [9]. We aggregate opinions on features using summaries of different products so that consumer opinions can be compared to competing products. Figure 4 shows a visualization of customer opinions of different entities.

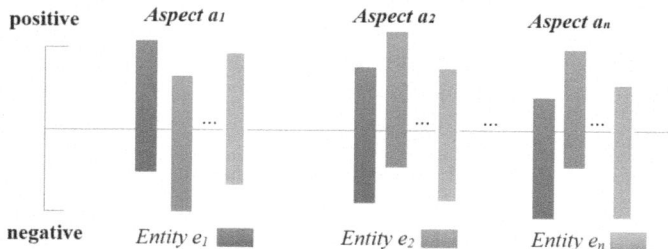

Fig. 4. Visualization of opinions on different entities

Where, $e = \{e_1, e_2, \ldots, e_n\}$ is the set of entities that are the names of the products corresponding to the set $a = \{a_1, a_2, \ldots, a_n\}$ aspects of the entity e. Each entity can also have many different aspects. The entity opinion or aspect of the entity is positive or negative.

In each of the different colored bars, Fig. 4 shows the percentage of reviews that express a negative or positive opinion on a horizontal axis about a certain aspect. With the above opinion summarization, it is very convenient for customers to visually observe to know the strengths and weaknesses of each product to make product selection decisions.

8 Experiment

8.1 Data Used

In this experiment, we used two datasets YelpHotelData and YelpResData [6] as the basis for training the proposed model.

The YelpHotelData dataset is a database containing content related to hotel opinions, hotel reviewers, and hotel reviews. This dataset has been extracted, and processed into structured data containing important information from reviews related to aspects of the hotel such as credit card payment services *(AcceptsCreditCards)*, in relation to the price service *(PriceRange)*, and the hotel's Wi-Fi facility *(WiFi)*, opinion holder and time of review are also listed in the data. The details of the YelpHotelData dataset are described in detail in Tables 2 and 3.

The YelpResData dataset contains content related to opinion restaurants, restaurant reviewers, and restaurant reviews. Similar to the YelpHotelData dataset, this dataset contains more restaurant aspects than the YelpHotelData dataset. The details of the YelpHotelData dataset are described in detail in Tables 4 and 5.

Table 2. The details of tables of the YelpHotelData dataset

Tables	Total columns	Total rows
Hotel	13	283086
Reviewer	13	5123
Review	10	688329

Table 3. The details of columns of the YelpHotelData dataset

Hotel	Reviewer	Review
HotelID	ReviewID	Date
Name	Name	ReviewID
Location	Location	ReviewerID
ReviewCount	YelpJoinDate	ReviewContent
Rating	FriendCount	Rating
Categories	ReviewCount	UsefulCount
Address	FirstCount	CoolCount
AcceptsCreditCards	UsefulCount	FunnyCount
PriceRange	CoolCount	Flagged
WiFi	FunnyCount	HotelID
Website	ComplimentCount	
PhoneNumber	TipCount	
FilReviewCount	fanCount	

Table 4. The details of tables of the YelpResData dataset

Tables	Total columns	Total rows
Restaurant	30	242652
Reviewer	13	16941
Review	10	788471

Table 5. The details of columns of the YelpResData dataset

Restaurant	Restaurant	Reviewer	Review
RestaurantID	Delivery	ReviewID	Date
Name	Takeout	Name	ReviewID
Location	WaiterService	Location	ReviewerID
ReviewCount	OutdoorSeating	YelpJoinDate	ReviewContent
Rating	WiFi	FriendCount	Rating
Categories	GoodFor	ReviewCount	UsefulCount
Address	Alcohol	FirstCount	CoolCount
Hours	NoiselLevel	UsefulCount	FunnyCount
GoodforKids	Ambience	CoolCount	Flagged
AcceptsCreditCards	HasTV	FunnyCount	RestaurantID
Parking	Caters	ComplimentCount	
Attire	WheelchairAccessible	TipCount	
GoodforGroups	WebSite	FanCount	
PriceRange	PhoneNumber		
TakesReservations	FilReviewCount		

8.2 Preprocessing

The YelpHotelData dataset is preprocessed from the original data converting the data into a five-component structure of the opinion as *(Hotel names, Hotel aspects, Opinion ratings, Hotel reviewers, Reviewer times)* corresponding to the opinion quintuple (e, a, s, h, t) have been suggested above. The value of the five components e, a, s, h, t is converted to a numeric value. The details of data conversion results have 882,474 opinion quintuples are reported in Table 6.

Table 6. The details of hotel data conversion

The features of hotel review	Different values of features
Hotel names	from 1 to 123461
Hotel aspects	from 1 to 3
Hotel reviewers	from 1 to 4596
Reviewer times	from 1 to 4382
Opinion ratings	from 1 to 5

Similarly, the YelpResData dataset is preprocessed into a five-component structure of the opinion as *(Restaurant names, Restaurant aspects, Opinion ratings, Restaurant*

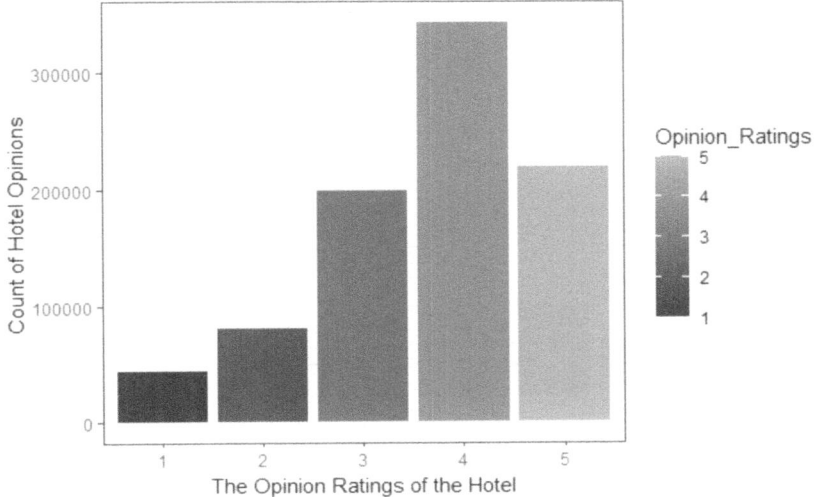

Fig. 5. Distribution of opinion ratings about the hotel

reviewers, Restaurant times). The distribution of opinion ratings about the restaurant in Fig. 6. We explored nineteen aspects that influence the quality of the restaurant such as GoodforKids, AcceptsCreditCards, Parking, Attire, GoodforGroups, PriceRange, TakesReservations, Delivery, Takeout, WaiterService, OutdoorSeating, WiFi, GoodFor, Alcohol, NoiselLevel, HasTV, Ambience, Caters, and WheelchairAccessible. The results of the initial data conversion have 1,479,1500 sets of opinion quintuples (Fig. 5).

Table 7. The details of restaurant data conversion

The features of restaurant	Different values of features
Restaurant names	from 1 to 184167
Restaurant aspects	from 1 to 19
Restaurant reviewers	from 1 to 12943
Restaurant times	from 1 to 4541
Opinion ratings	from 1 to 5

The ratings of the opinions of the reviews from 1 to 5 stars correspond to the five levels of opinion polarization as "very negative", "negative", "neutral", "positive", and "very positive". In the proposed model experiment, we propose to polarize opinions at two levels "negative" and "positive". Therefore, from the five polarizing perspectives we convert to two opinions polarity by combining the "very negative" and "negative" polarizing into "negative"; "positive", and "very positive" to "positive" and remove the polarization of the "neutral" opinion. The hotel opinion rating conversion results have 560,769 negative samples and 321,705 positive samples in Fig. 7. The restaurant

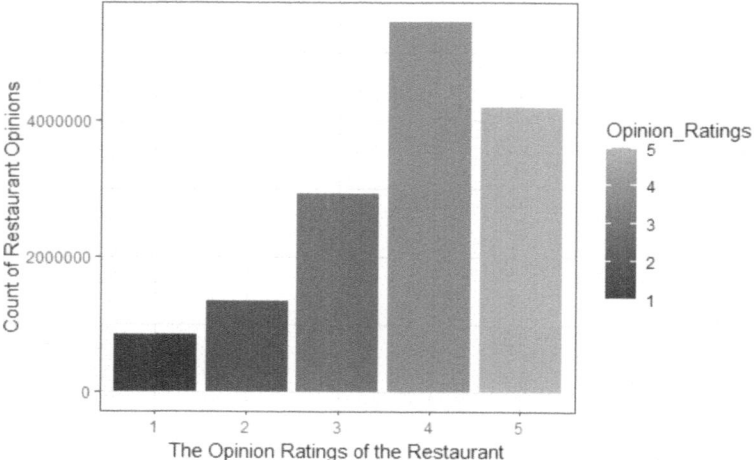

Fig. 6. Distribution of opinion ratings about the restaurant

opinion rating conversion results have 5,140,963 negative samples and 9,650,537 positive samples in Fig. 8 (Table 7).

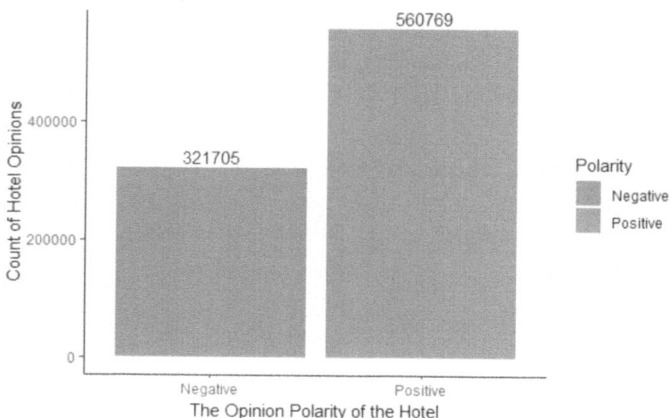

Fig. 7. The distribution of hotel opinions

8.3   Tool Used

In order to conduct experiments, we used R language to build the proposed model and integrated some main libraries into the model such as library(celltrackR) [10], library(stacks), library(randomForest), library(h2o), library(dplyr), and several support libraries for data processing and visualization.

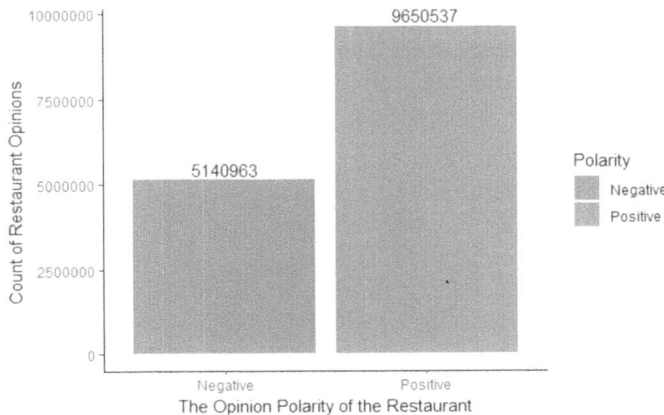

Fig. 8. The distribution of restaurant opinions

8.4   Scenario 1. Manifold Forests for Hotel Opinion.

The proposed model is trained on the YelpHotelData dataset that has been preprocessed to opinion quintubles and converted to feature vector space as input for training the model based on the following steps:

First, is to split the initial data set into a training dataset, validation dataset, and test dataset according to the following ratio tương ứng là 60%, 20%, 20%. The opinion training dataset is omitted from the labeling to perform clustering. As a result, three clusters are created.

Running three random forest models on 3 clusters such as the random forest 1 (RF1), the random forest 2 (RF2), and random forest 3 (RF3) with cross-validation method (k-folds = 5).

Get prediction results of 3 models RF1, RF2, RF3 from valid set (20%).

Build a random forest (RF) model to stack from the prediction results of three models such as RF1, RF2, RF3 to form a manifold forest (MF).

Comparing the accuracy of each model with the manifold forest (MF) on the test data set.

The performance in Table 8 of the manifold forest model on the hotel training set for accuracy, precision, recall, and F1 measures are 94%, 90%, 70%, and 79%, respectively. The testing set is 78%, 74%, 62%, and 79% for accuracy, precision, recall, and F1, respectively.

Table 8. Evaluate manifold forests for hotel opinion dataset

Hotel dataset	Accuracy	Precision	Recall	F1
Training set	0.94	0.90	0.70	0.79
Testing set	0.78	0.74	0.62	0.67

The comparison results are shown in Fig. 9. We can see the performance for each individual model. The accuracy of the random forest 3 (RF 3) for the training set is 90%,which is the lowest. The accuracy of RF 1, and RF 2 are 91%, and 92% respectively. The accuracy of MF is 94% for the training set, which is the highest.

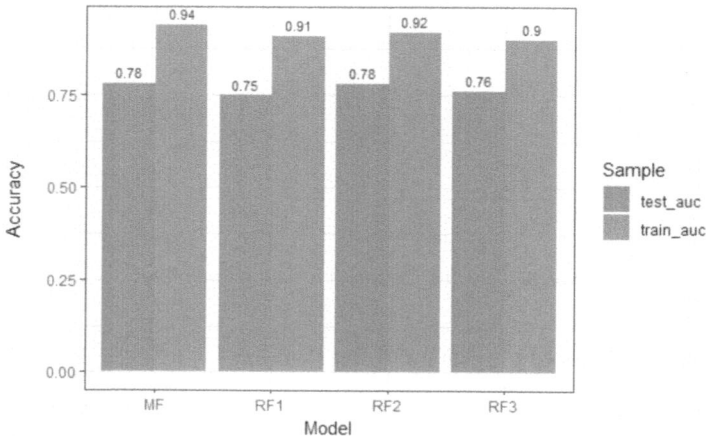

Fig. 9. The accuracy of the random forests 1, the random forests 2, the random forests 3, and manifold forests (MF) for the hotel opinion dataset.

8.5 Scenario 2. Manifold Forest for Restaurant Opinion.

Similar to scenario 1, we built the manifold forest for the restaurant opinion dataset by applying the following steps:

First, we have divided data set into a training dataset, validation dataset, and test dataset with ratios of 60%, 20%, 20% respectively. Then perform clustering on the unlabeled training dataset. As a result, four clusters are created. We build four independent random forests including RF1, RF2, RF3, and RF4 respectively on four clusters with the cross-validation method (k-folds = 10).

Get prediction results of four models RF1, RF2, RF3, and RF4 from the valid set (20%).

Applying random forest (RF) to stack from the prediction results of four models such as RF1, RF2, RF3, and RF4 to form a manifold forest (MF).

The manifold forests on the restaurant opinion training set in Table 9 have a performance of 95%, 92%, 71%, and 80% for accuracy, precision, recall, and F1 measures, respectively. The testing set is 82%, 75%, 67%, and 71%, for accuracy, precision, recall, and F1 measures, respectively

The accuracy of RF 1, RF 2, RF 3 are 93%, 94%, 92% for the training set, respectively. In which, the accuracy of RF 3 is the lowest. The accuracy of MF is 95%, which is the highest for the training set in Fig. 10.

Table 9. Evaluate manifold forests for the restaurant opinion dataset

Restaurant dataset	Accuracy	Precision	Recall	F1
Training set	0.95	0.92	0.71	0.80
Testing set	0.82	0.75	0.67	0.71

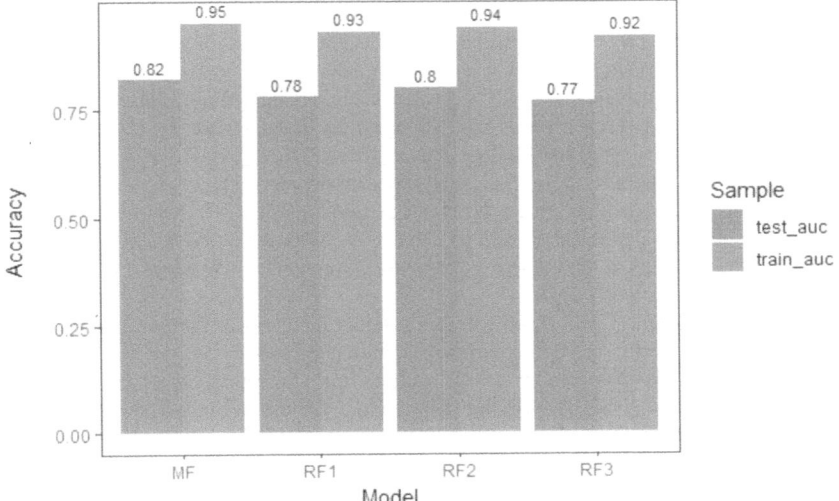

Fig. 10. The accuracy of forests for the restaurant opinion dataset

9 Conclusion

In this study, we propose to build an opinion mining model with the manifold forests approach to solve the problem of clustering opinions according to the affinity between different clusters of opinion in large-scale opinion data. In particular, we focus on building a random forest model on the clusters to determine the affinity of the opinion data and aggregate the forests by ensemble methods to identify the opinion classification in reviews as either positive or negative. We conducted experiments using a dataset of hotel and restaurant reviews. The results show that manifold forests can accurately estimate opinion classification, and the use of ensemble learning, such as the stacking algorithm, achieves the best results. This means that ensemble learning can improve the accuracy and efficiency of opinion classification in hotel and restaurant reviews. This finding could be useful for opinion mining to choose an effective model.

Acknowledgments. The authors gratefully acknowledge the support of distinguished professor Bing Liu, Department of Computer Science University of Illinois at Chicago (UIC) for sharing the dataset to conduct this study.

References

1. Li, Z., Chen, Y., LeCun, Y., Sommer, F.T.: Neural manifold clustering and embedding. In: Proceedings of the 10th International Conference on Learning Representations (ICLR) (2022)
2. Shiebler, D.: Functorial manifold learning and overlapping clustering. In: Proceedings of Machine Learning Research, vol 132, pp.1–20 (2021)
3. Perry, R., Tomita, T. M., Patsolic, J., Falk, B., Vogelstein, J.T.: Manifold forests: closing the gap on neural networks. In: Proceedings of the 8th International Conference on Learning Representations (ICLR) (2019)
4. Souvenir, R., Pless, R.: Manifold clustering. In: Proceedings of the 10th International Conference on Computer Vision (ICCV), pp. 648–653 (2005)
5. Ester, M., Kriegel, H., Sander, J., Xu, X.: A density-based algorithm for discovering clusters in large spatial databases with noise. In: Proceedings of 2nd International Conference on Knowledge Discovery and Data Mining (KDD), ACM, 226–231 (2016)
6. Bordoloi, M., Biswas, S.K.: Sentiment analysis: a survey on design framework, applications and future scopes. Artif. Intell. Rev. (2023). https://doi.org/10.1007/s10462-023-10442-2
7. Mukherjee, A., Venkataraman, V., Liu B., Glance N.: What yelp fake review filter might be doing. In: Proceedings of the International AAAI Conference on Weblogs and Social Media (ICWSM), Boston, USA (2013)
8. Zhai, Z., Liu, B., Xu, H., Jia, P.: Clustering product features for opinion mining. In: Proceedings of the Fourth ACM International Conference on Web Search and Data Mining (WSDM'11), pp. 347–354 (2011)
9. Liu, B.: Sentiment Analysis: Mining Sentiments, Opinions, and Emotions. 2nd edn. Cambridge University Press, Cambridge (2020)
10. Breiman, L.: Random forests. Mach. Learn. **45**, 5–32 (2001)
11. Tran, P.Q., Trieu, N.T., Dao, N.V., Nguyen, H.T., Huynh, H.X.: Effective opinion words extraction for food reviews classification. Int. J. Adv. Comput. Sci. Appl. **11**(7), 421–426 (2020)
12. Gautam K.: Ensemble Methods for Machine Learning. Manning Publications Co., Shelter Island, NY (2023)
13. Wortel, I.M.N., Liu, A.Y., Dannenberg, K., Berry, J.C., Miller, M.J., Textor, J .: CelltrackR: an R package for fast and flexible analysis of immune cell migration data. ImmunoInform. 1–2 (2021). https://ingewortel.github.io/celltrackR/
14. Tran, P.Q., Nguyen, H.T., Le, H.M.T., Huynh, H.X.: Ensemble learning for mining opinions on food reviews. In proceedings of the International Conference on Context-Aware Systems and Applications(ICCASA 2021), pp 56–70 (2021)
15. Powers, D.M.W.: Evaluation: from precision, recall and f-measure to roc., informedness, markedness and correlation. J. Mach. Learn. Technol. **2**, 37–63 (2011)
16. Tran, P.Q., Ha, D.N.L., Le, H.T.M., Huynh, H.X: Opinion mining with density forests. J. EAI Endorsed Trans. Context-aware Syst. Appl. **9**(1), (2023)

Systematic Review of Smart Robotic Manufacturing in the Context of Industry 4.0

Lu Anh Duy Phan[1,2] and Ha Quang Thinh Ngo[1,2](✉)

[1] Department of Mechatronics, Faculty of Mechanical Engineering, Ho Chi Minh City University of Technology (HCMUT), 268 Ly Thuong Kiet Street, District 10, Ho Chi Minh, Vietnam
[2] Vietnam National University-Ho Chi Minh City (VNU-HCM), Linh Trung Ward, Thu Duc City, Ho Chi Minh, Vietnam
nhqthinh@hcmut.edu.vn

Abstract. In the rapidly evolving landscape of industrial revolution, smart robotic manufacturing has emerged as a game-changing phenomenon, revolutionizing traditional production processes, and unlocking unprecedented levels of productivity and safety. At the core of this transformative paradigm lies the seamless integration of artificial intelligence (AI) models, empowering autonomous robotic systems to carry out complex tasks with unparalleled precision and efficiency. This systematic review endeavors to explore the synergistic relationship between smart robotic manufacturing and AI technologies, delving into the various advancements, challenges, and potential implications for industrial sectors. By shedding light on the cutting-edge innovations and practical insights, this work aims to provide a valuable resource for researchers, industrialists, and policymakers seeking to leverage the transformative potential of AI-driven smart robotic manufacturing in the era of Industry 4.0.

Keywords: Intelligent Robot · Collaborative Robot · Smart Manufacturing · Predictive Model · Advanced Interaction

1 Introduction

In recent years, robotics has witnessed remarkable advancements in various domains, including manufacturing and production. Integrating machine learning techniques, particularly reinforcement learning, with robotics has been pivotal in developing intelligent robotic manufacturing systems. These systems aim to enhance the efficiency of industrial processes, flexibility, and adaptability by enabling robots to learn from their environment and make intelligent decisions.

This analysis focuses on applying reinforcement learning techniques in two specific cases of manufacturing: robotic manipulation [1] and collaborative robotics [2]. Both approaches have revolutionized traditional manufacturing practices by introducing a new generation of robots to perform complex tasks with precision and agility. In some

© ICST Institute for Computer Sciences, Social Informatics and Telecommunications Engineering 2024
Published by Springer Nature Switzerland AG 2024. All Rights Reserved
P. Cong Vinh and N. Thanh Tung (Eds.): ICCASA 2023, LNICST 579, pp. 19–42, 2024.
https://doi.org/10.1007/978-3-031-58878-5_2

factories with high rate of automation, only robots appear in their workplace and reach to self-assessment and decision-making. Though, in the lower level of automation, it requires that human co-exists and supports to manipulate.

In the high level, robotic manipulation is a fundamental aspect of manufacturing, involving precisely handling and manipulating objects. It is very crucial to manipulate robot based on our desired missions. Formerly, traditional robotic manipulation systems were limited in adapting to object shape, size, and position variations. However, with the advent of learning technologies, robots can now acquire new skills and autonomously adjust to changes in their environment. Several methods such reinforcement learning [3], deep learning [4], and computer vision [5] have been instrumental in enhancing the capability of robots to grasp, manipulate, and assemble objects with better accuracy and efficiency.

For lower level, collaborative robotics, on the other hand, focuses on the interaction between robots and workers. Henceforth, robots and humans worked in separate, isolated spaces in traditional manufacturing settings due to safety concerns. Nevertheless, collaborative robots, also known as cobots, are designed to work alongside humans, facilitating teamwork and cooperation. This approach involves robots learning to understand human gestures, intentions, and actions, enabling them to aid, share tasks, and ensure safe and efficient collaboration. Machine learning algorithms, such as human-robot interaction models and motion planning techniques, have been instrumental in developing effective collaborative robotic systems that enhance productivity and workplace safety.

The remainder of this study is arranged as follows. In Sect. 2, it provides an introduction to the background and concepts related to smart robotic manufacturing, machine learning, and robot learning. Section 3 delves into robotic manipulation, exploring the advancements and techniques used in this domain. Section 4 discusses human-robot collaboration in manufacturing applications, highlighting the key aspects and real-world circumstances. Finally, Sect. 5 presents open problems and future research directions for potential exploration and advancement in this field.

2 Background and Concepts

This section provides an overview of the background and fundamental concepts relevant to the robot learning scheme toward smart robotic manufacturing. Understanding these concepts is essential for comprehending the advancements and applications discussed in subsequent sections.

2.1 Smart Manufacturing

Smart manufacturing represents a modern and highly integrated manufacturing approach that combines the latest advancements in information technology, including the Internet of Things (IoT), cloud computing, and artificial intelligence (AI), with cutting-edge manufacturing processes [6]. By utilizing these technologies, smart manufacturing seeks to improve the productivity of production processes through autonomous perception, improved decision-making, and accurate execution, ushering in a new era of vitality for manufacturing [7]. This intelligent paradigm presents exciting opportunities for the global manufacturing industry, paving the way for more flexible, adaptive, and personalized production processes. The transformation and advancement of smart manufacturing hold significant implications for the overall development of manufacturing on a global scale.

2.2 Machine Learning

Machine learning is a multidisciplinary field that combines computation and statistics with connections to information theory, signal processing, algorithms, control theory, and optimization theory [8]. Machine learning has emerged as an exceptionally captivating study area in artificial intelligence. The fundamental concept of machine learning revolves around constructing models that can approximate data-based functions. Once trained on available data, these models can then be utilized to approximate and predict outcomes for new data instances. This process of transitioning from an initially weak model to a more robust one by leveraging the available data is termed "learning," analogous to the learning capabilities and processes observed in living organisms. Given that this process is predominantly accomplished by machines, i.e. computers, the term such machine learning was coined to represent this field of study.

2.3 Robot Learning

Robot learning refers to integrating various machine learning technologies within the field of robotics [9]. It specifically focuses on applying machine learning techniques to enable robots to learn and make informed decisions. Unlike traditional machine learning, robot learning emphasizes generating actions as output while perceiving and understanding the environment as input. For example, deep learning techniques enhance a robot's ability to navigate and interact with unstructured environments effectively. On the other hand, reinforcement learning provides formal frameworks to govern machine behaviors and decision-making processes. By synthesizing these machine learning technologies, robot learning enables robots to acquire knowledge, adapt to their surroundings, and engage in intelligent actions within their environment.

2.4 Reinforcement Learning

Reinforcement learning (RL) is a specialized branch within machine learning dedicated to instructing agents in making a series of decisions in an environment with the objective of maximizing a cumulative reward. RL is inspired by behavioral psychology, where an agent learns through trial and error by interacting with its environment. As shown in Fig. 1, an agent interacts with an environment and takes actions based on its current state. The working environment provides feedback to the agent in the form of rewards, which indicate the desirability of the actions of an agent. The goal of agent is to learn a policy a mapping from states to actions that maximizes the expected cumulative reward over time.

A lot of recent research choose to categorize RL algorithms into model-based approach and model-free approach, despite the fact that this is a challenging task given their

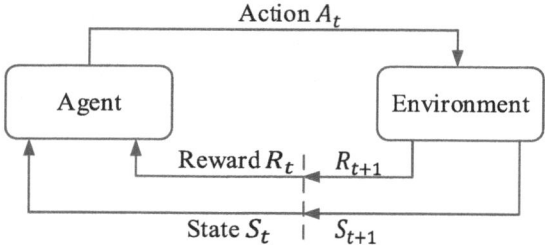

Fig. 1. Diagram for Learning Framework of Reinforcement Learning

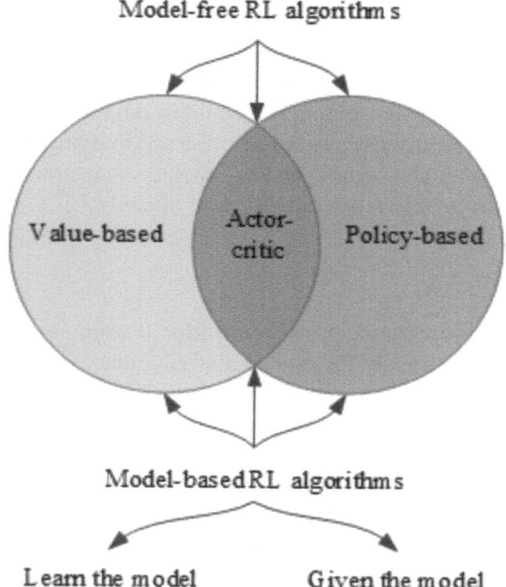

Fig. 2. Description of classifications for reinforcement learning algorithms [10].

extensive modularity. In Fig. 2, these researches, in turn, are divided into three main types: value-based, policy-based and a hybrid method called actor-critic algorithms [10].

A. Value-based RL.

Value-based algorithms generate the value function for each state or state-action pair until their values converge rather than storing any explicit policies. Instead, they achieve this via temporal difference learning. This makes it possible to decrease the variation in estimates of the expected returns, but it also necessitates a time-consuming optimization process. By acting greedily (choosing the action with the best value) on the calculated function, the optimum policy may be easily deduced from the value function. In this study, some of the algorithms of interest are as follows:

Q-Learning. Q-Learning [11] is one of the most used RL algorithms that learns the optimal action-value function (Q-function) through iterative updates. It uses the Bellman equation to update Q-values based on observed rewards and estimated future rewards.

Deep Q-Learning (DQN). DQN [12] is an extension of Q-learning that incorporates deep neural networks to approximate the Q-function. It uses experience replay to store and sample experiences, improving data efficiency and stabilizing learning.

Double Deep Q-Learning (Double DQN). Double DQN [13] addresses the issue of overestimation bias in Q-learning. It uses two sets of Q-values to decouple action selection and value estimation, reducing the tendency to overestimate the values of actions.

Dueling Deep Q-Learning (Dueling DQN). Dueling DQN [14] separates the estimation of the value and advantage functions, allowing the agent to learn the value of being in a particular state independently of the advantage of each action. This architecture enables better learning efficiency and generalization.

B. Policy-based RL.

In contrast to value-based algorithms, policy-based algorithms directly create the policy responsible for assigning the best possible action to each state. This parametrized function describing the approach is kept in memory throughout the learning process. These algorithms thereby improve the policy without relying on value function estimations. With this method, they can develop a smooth range of activities, but the variability may rise. Either gradient-based or gradient-free parameter estimation techniques can derive these algorithms. Several significant algorithms in this study include:

Vanilla Policy Gradient (VPG). VPG method is a direct application of the Policy Gradient Theorem [15]. The main idea behind VPG is to update the policy of agent in the direction that increases the expected reward. It achieves this by estimating the gradient of the expected reward with respect to the policy parameters and using this gradient to update the policy.

Trust Region Policy Optimization (TRPO). TRPO [16] is a policy optimization algorithm that seeks to improve the policy while ensuring small policy updates to maintain stability. It employs a trust region constraint to bind the maximum policy update step based on the Kullback-Leibler divergence.

Proximal Policy Optimization (PPO). PPO [17] is an advanced reinforcement learning algorithm designed to overcome the computational overhead of TRPO. While TRPO offers stability and convergence guarantees, it can be computationally expensive due to the need for solving complex constrained optimization problems at each update.

C. Actor-critic algorithms.

Actor-critic algorithms [18] represent a fusion of value-based and policy-based techniques within reinforcement learning. In this methodology, the 'actor', represented by a policy network, suggests actions for a specific state. In contrast, the 'critic', embodied by a value network, assesses these actions within the context of state-action pairs. By employing the Bellman equation, the critic learns the Q-function, and the actor is updated based on the Q-function to train the policy. Through this dual mechanism, the actor-critic approach harnesses the respective advantages of both value-based and policy-based methods. Up to date, the actor-critic based algorithms are mostly used and the classical algorithms include:

Advantage Actor-Critic (A2C). A2C [19] is an actor-critic algorithm that combines the advantages of both policy-based and value-based methods. It maintains a policy (actor) and a value function (critic) to estimate the advantage of each action. The policy is updated using the advantage as a baseline.

Asynchronous Advantage Actor-Critic (A3C). A3C [19] is an extension of A2C that parallelizes the learning process by using multiple agents that interact with different environment instances. Each agent updates the shared policy and value function asynchronously, improving sample efficiency.

Deep Deterministic Policy Gradient (DDPG). DDPG [20] is a reinforcement learning technique combining both deep Q-learning and deterministic policy gradients (DPG) [21]. DDPG has two main components: the actor-network and the critic network. The actor-network learns the policy, mapping states to actions, while the critic network estimates the Q-value of state-action pairs. These networks are usually implemented as deep neural networks, allowing them to approximate complex functions.

Twin Delayed Deep Deterministic Policy Gradient (TD3). TD3 [22] is an advanced variant of the Deep Deterministic Policy Gradient (DDPG) algorithm. TD3 was proposed to improve the stability and sample efficiency of DDPG in continuous action space reinforcement learning problems.

Soft Actor-Critic (SAC). SAC [23] is a state-of-the-art deep reinforcement learning algorithm designed for environments with continuous action spaces. It is based on the actor-critic architecture and has been proven to be highly effective and stable in various continuous control tasks. SAC has demonstrated impressive performance on various challenging continuous control tasks, such as robotic control, locomotion, and dexterous manipulation.

2.5   Network Architecture

Neural networks are effective function approximators in deep reinforcement learning, especially when the state or action space is too large to be fully known. Among the neural network architectures commonly employed in deep RL, the multilayer perceptron (MLP) is prevalent. An MLP comprises an input layer, a hidden layer, and an output layer. Unlike the input nodes, neurons in the hidden and output layers employ nonlinear activation functions. MLPs are trained through backpropagation, a supervised learning technique. However, MLPs possess a drawback in that they are fully connected, establishing connections between every perceptron. Consequently, this can result in many parameters and redundant information in high-dimensional spaces, making them inefficient.

A. Convolutional Neural Network.

Convolutional neural network (CNN) [24] is a type of artificial neural network specifically designed for processing and analyzing visual data, such as images and videos.

CNNs have proven to be highly effective in tasks related to computer vision, including image classification, object detection, and image segmentation. The key feature of CNNs is their ability to automatically learn and extract hierarchical features from input data through a series of convolutional layers. These layers use convolution operations to scan and filter the input data, gradually capturing patterns and features of increasing complexity. CNNs typically consist of multiple layers, including convolutional layers, pooling layers for down-sampling, and fully connected layers for classification or regression tasks. In a reinforcement learning algorithm, the output of the CNN can be used as the input for a value or policy function.

B. Recurrent Neural Network.
A recurrent neural network (RNN) [25] is a type of artificial neural network designed to process sequential data by maintaining internal memory. Unlike traditional feedforward neural networks, which process data in a one-way direction, RNNs have connections that loop back on themselves, allowing them to capture patterns and dependencies in sequences. The key feature of an RNN is its ability to retain information from previous time steps and use it to influence the processing of the current input. This makes RNNs particularly well-suited for sequences involving natural language processing, speech recognition, time series analysis, and more.

C. Graph Neural Network.
Graph Neural Network (GNN) [26] is a type of neural network designed to process and make predictions on graph-structured data. Graphs are mathematical structures that consist of nodes (vertices) connected by edges (links), and they are used to model relationships or interactions between different entities. GNN can be used for various tasks, including node classification, link prediction, graph classification, and recommendation systems, among others. Some popular GNN architectures include Graph Convolutional Networks (GCN) [27], GraphSAGE [28], and Graph Attention Networks (GATs) [29]. These architectures have been successfully applied to a wide range of domains, such as social networks, bioinformatics, chemistry, and knowledge graphs.

3 Deep Reinforcement Learning for Robotic Manipulation

This section explores how Deep Reinforcement Learning (Deep RL) techniques enhance robotic manipulation. From grasping techniques to gripper designs and theoretical simulation, these advancements empower robots with improved precision, dexterity, and adaptability in manipulation.

3.1 Robotic Grasping

Robotic grasping has improved Through various tactics meant to increase the effectiveness and adaptability of object manipulation. These include sole-grasping rules, in which specific algorithms precisely guide robots to acquire objects in challenging circumstances. Additionally, suction-based grasping methods have increased the variety of things that robots may handle by enabling secure grabbing by establishing a vacuum seal.

Thanks to multifunctional grippers, robots are now adaptable and can change their grip to suit different object shapes and sizes. In addition to these methods, "two-action synergy" has been developed, which entails the coordinated performance of several actions, such as grasping and pushing, for improved grasping performance.

Combining these robotic grasping approaches with machine learning can potentially have transformational effects on various automation and robotics-related businesses, especially when combined with the development of machine learning techniques, particularly reinforcement learning. For a comprehensive description, Table 1 depicts the literature on Deep RL in robotic grasping action.

A. Sole-grasping.

This policy is a technique for training a robot to only utilize the grasping action to catch objects when no other action (such as pushing, moving, and poking) are involved. Numerous research studies have focused on applying reinforcement learning to grasping individual or cluttered objects. For instance, there are studies [30, 31] in the case of learning to grasp individual objects. Investigators in [30] introduced a reinforcement learning model that utilizes a strategy search algorithm, demonstrating remarkable robustness in the generalization from simple to complex object shapes. However, the current form of guided policy search (GPS) faces limitations in its applicability to sequential multitask learning scenarios due to its batch-style training requirement. In another related study, developers in [31] concentrates on addressing the problem of chin-grasping poses in 3D space using 3D point clouds as inputs for the model. The research results showed promising simulation outcomes, and the simulation data can be further utilized for real-world applications.

Additionally, accurate grasping is a challenging problem with significant potential for applications in manufacturing. Researchers in [32] developed an innovative approach for training a robot to perform pick-and-place tasks using self-supervised learning, without relying on an object model. They combined two techniques, namely, robot learning of primitives estimated by fully convolutional networks (FCNs) and one-shot imitation learning (IL). To achieve precise pick-and-place actions without an object model, they formulated the place reward as a contrastive loss between real-world measurements and a task-specific noise distribution. The results showed great promise in terms of accuracy. However, the process required a substantial amount of time for exploring behaviors, which limited its efficiency for industrial applications. In general, this study represents a significant step forward in developing robotic capabilities for pick-and-place tasks without the need for explicit object models. A highly innovative grasping strategy is proposed in a study [33]. TossingBot, a robot system capable of accurately tossing various objects to designated target positions, showcases the potential of this approach. The authors introduce an end-to-end approach that simultaneously learns to infer control parameters for grasping and throwing by iteratively testing and adjusting based on images of objects within a container. This self-supervised learning process enables the system to identify optimal grasping positions that result in consistent and predictable throws. To simplify the throwing task, the system focuses on predicting the release velocity alone. The release velocity is determined using a physics-based controller and further refined based on the residual estimate obtained from the neural network. The integration of image-based learning and physics-based control in TossingBot demonstrates promising results in enhancing the ability of a robot to handle a wide array of objects and execute precise throwing actions.

Table 1. Literature of Deep Reinforcement Learning in Robotic Grasping

Author(s)	Publication year	Action	Gripper type	Sim package	Sim/Real-world	Method
Beltran-Hernandez et al. [30]	2019	Grasping	Parallel-jaw	Gazebo	Sim	CNN + Guided Policy Search
Mousavian et al. [31]	2019	Grasping	Parallel-jaw	FleX	Sim/Real	Point-Net + +
Berscheid et al. [32]	2020	Grasping and placing	Parallel-jaw	N/A	Real	FCNNs + Q-learning + one-shot imitation learning
Zeng et al. [33]	2020	Grasping and throwing	Two fingers	Bullet	Sim/Real	ResNet-FCNs + Q-learning
Shao et al. [34]	2019	Grasping	Suction	V-REF	Sim	Resnet with U-net (CNN) + Q-learning
Zakka et al. [35]	2020	Grasping and placing	Suction	N/A	Real	FCN ResNet + Q-learning
Cao et al. [36]	2022	Grasping	Suction	V-REF	Sim	A3C
Zeng et al. [37]	2022	Grasping	Two fingers, suction	N/A	Real	ResNet-FCNs + Q-learning
Zeng et al. [38]	2018	Grasping and pushing	Two fingers	V-REF	Sim/Real	DenseNet-FCNs + Q-learning
Ren et al. [39]	2021	Grasping and moving	Two fingers	V-REF	Sim/Real	Duelling DDQN
Tang et al. [40]	2021	Grasping and pushing	Three fingers	V-REF	Sim/Real	DenseNet-FCNs + Q-learning
Zhang et al. [41]	2023	Grasping and pushing	Two fingers	Isaac Gym	Sim/Real	PPO
Berscheid et al. [42]	2019	Shifting and grasping	Parallel-jaw	N/A	Real	Deep Q-learning

(continued)

Table 1. (*continued*)

Author(s)	Publication year	Action	Gripper type	Sim package	Sim/Real-world	Method
Hundt et al. [43]	2020	Pushing, Grasping and Placing	Two fingers	V-REF	Sim/Real	DenseNet-FCNs + Q-learning
Yang et al. [44]	2020	Grasping and pushing	Parallel-jaw	V-REF	Sim/Real	FCNs + Q-learning
Xu et al. [45]	2021	Grasping and pushing	Two fingers	V-REF	Sim/Real	DenseNet-FCNs + Q-learning
Huang et al. [46]	2021	Grasping and pushing	Two fingers	Bullet	Sim/Real	DQN + MCTS + DIPN
Chebotar et al. [47]	2021	Grasping and placing	Two fingers	N/A	Sim/Real	Q-learning
Ren et al. [48]	2022	Grasping and pushing	Two fingers	V-REF	Sim/Real	DenseNet-FCNs + Q-learning
Novkovic et al. [49]	2020	Grasping and pushing	Parallel-jaw	Bullet	Sim/Real	PPO
Chen et al. [50]	2020	Grasping and pushing	Parallel-jaw	MuJoCo	Sim	TD3

B. Suction-Based Grasping

Suction grasping is another mechanism strategy that has been increasingly utilized for performing object manipulation in dense cluttered environments. This approach involves using suction-based grippers or end-effectors that generate a vacuum force to firmly attach to the surface of an object. By creating a secure suction grip, robots can pick up and manipulate objects even in scenarios where traditional grasping mechanisms might struggle due to clutter or irregular shapes. Suction grasping proves to be particularly effective in industries such as logistics, warehousing, and agriculture, where objects may be randomly arranged or piled up in confined spaces. The ability to handle objects in dense clutter allows robots to operate efficiently, improving automation capabilities and expanding their scope of applications in real-world scenarios.

In [34], researchers introduced the concept of suction grasp as a viable alternative for object manipulation in cluttered environments, aiming to mitigate potential failure situations resulting from the combination of pushing and grasping actions. Their approach involved utilizing deep reinforcement learning, using techniques such as Q-learning with ResNet and the U-net structure. Their method faced a limitation in that the suction grasp points were randomly predicted, leading to difficulties in accurately identifying grasp points, especially in cluttered environments. To enhance the effectiveness of such frameworks, there is a need to incorporate more diversity in the shapes of objects during both training and testing phases. Moreover, their reported results were solely based on successful outcomes in simulated environments using CoppeliaSim (V-REP), raising the need for additional real-world validation To demonstrate a method for pick-and-place, using matching network [35] computes dense visual descriptors to associate picking actions to placing actions. Their framework can easily be generalized to new objects and kits. Yet, due to the only processing 2D rotations and some assumptions that objects are face-down, it would be interesting to explore a more complex action representation for 3D assembly. In another related study [36], researchers proposed using an Actor-Critic algorithm, A3C, as the reinforcement learning method to train the picking policy network for executing grasping tasks in cluttered areas. This opens exciting possibilities for improving robotic manipulation in complex scenarios.

C. Multifunctional Gripper-Based Grasping
Recently, another intriguing mechanism gaining attention in research studies is the training of reinforcement learning algorithms to coordinate the execution of grip and suction grasps. This approach involves equipping the robotic arm with a gripper capable of both finger gripping and suction-cup functions. By adopting this multifunctional gripper design, researchers can leverage the advantages of both gripping techniques in various scenarios. For instance, the finger-gripper excels at grasping objects in cluttered environments, overcoming the limitations of the suction-cup grasp, and vice versa. In one such study [37], the researchers proposed a method for robotic object manipulation, specifically pick-and-place tasks, by predicting both grip and suction affordances using the multifunctional gripper. Their approach employed a fully convolutional residual network to predict suction affordance for multi-view RGB-D images. A category-agnostic affordance prediction technique was then utilized to choose and execute one of four potential grasping primitive behaviors. However, it's worth noting that the inclusion of planar grasps in their learning approach might pose challenges due to arm movement restrictions. Furthermore, the strategy of 'Pick first, ask questions later' employed in their study may not be suitable for tasks requiring pre-determining the target object. This ongoing research in RL coordination of grip and suction grasps holds great promise for advancing robotic manipulation capabilities and overcoming complex real-world challenges.

D. Synergy of Two Primitive Actions
The synergy of two primitive actions refers to the powerful combination of two fundamental movements or behaviors that, when integrated, create a more complex and efficient action. In the context of robotics and artificial intelligence, primitive actions represent basic building blocks of behavior, such as grasping, pushing, reaching, or turning. By combining these individual actions in a coordinated manner, robots can perform more sophisticated tasks and adapt to dynamic environments effectively. This synergy enables robots to handle complex real-world scenarios that require a sequence of actions, making them more versatile and capable problem-solvers. The concept of combining primitive actions is essential for developing advanced robotic systems that can accomplish a wide range of tasks autonomously and adaptively, bringing us closer

to the realization of intelligent and adaptable robotic agents in various fields. As technology continues to evolve, harnessing the synergy of primitive actions opens up new frontiers for robotics research and applications, propelling us towards a future where robots seamlessly integrate into our daily lives, assisting us in a multitude of tasks and enriching our experiences.

In [38], scholars delved into exploring the synergistic relationship between pushing and grasping in the context of robotics. They aimed to achieve more stable and efficient results in densely cluttered environments by training deep end-to-end policies. Their approach was centered on the visual pushing-grasping (VPG) framework, where Q-learning was employed with the DenseNet pre-training model, specifically DenseNet-FCN, a fully connected network. However, it was noted that the VPG framework was primarily designed for target-agnostic tasks, necessitating re-projection before inputting the prediction network. One limitation observed was that the intrinsic pushing reward did not explicitly indicate whether a push would facilitate future grasping. Consequently, there were instances where the robot inadvertently pushed objects out of the workspace, leading to unnecessary actions and prolonged task execution times. Similar studies exploring the combination of pushing and grasping in the context of robotics have been of interest [39–41]. Developers [39] concentrate on the rapid acquisition of grasping and pre-grasping skills in goal-agnostic tasks. Their study aims to enhance performance and learning efficiency by introducing a mask function. This function plays a crucial role in guiding the robot's grasping behavior, enabling it to identify and focus on relevant object features during the learning process. By incorporating the mask function, the robot becomes more adept at grasping objects in diverse and unpredictable environments, ultimately leading to improved overall performance in goal-agnostic tasks. In [40] presented a novel method for collaborative pushing actions to aid in the process of grasping objects. Their approach employed Q-learning to learn a deterministic policy for both pushing and grasping. Interestingly, they did not assign any specific reward to pushing actions; instead, the agent received a reward only when the robot successfully completed the grasping task. Investigators in [41] develop a model-free Deep Reinforcement Learning framework to synergize pushing and grasping actions. The paper proposes an approach to tackle the challenge of handling objects in difficult positions. However, the push-to-wall method cannot effectively handle objects with bevels or hard flat sides.

Another approach to enhance object grasping in cluttered environments involves the technique of "shifting" objects, as discussed in [42], which involves putting a finger on top of the target object to increase grasp probabilities. The results of this approach demonstrate a high success rate and a capacity for generalization. Notably, the training process was conducted online without reliance on a simulation model. However, a critical aspect in improving the robotic bin grasping process is its ability to adapt when depth information is missing. This becomes particularly important because the availability of depth data from stereo cameras is often limited by factors such as shadows or reflective surfaces. Grasping objects near the edges or corners of tote boxes presents a notable challenge, and in scenarios where objects are densely packed together, the robot may encounter situations where suitable grasping options are not readily available.

Developers in [43] introduced a schedule outlining the Positive Task Framework (SPOT) and elaborated on the SPOT-Q RL algorithm. The SPOT framework serves the dual purpose of quantifying an agent's progress in multi-step tasks and offering crucial guidance with zero rewards, a masked action space, and situation removal. This framework can rapidly acquire policies that can generalize effectively from simulated environments to real-world scenarios. Nevertheless, it is worth noting that this approach places substantial demands on data resources and necessitates several iterative refinements to enhance its efficiency when applied to such tasks. Consequently, it is advisable to incorporate mechanisms for reactivity and failure recovery to counterbalance the precision loss that can occur due to policies trained in both simulated and real-world contexts.

In [44], researchers introduced a deep Q-learning approach for the purpose of grasping invisible objects, involving two distinct stages. The first stage revolves around determining the visibility of the target objects. If the thing is visible, the robot proceeds to the second stage and either performs a push or a grasp action. However, in cases where the target object remains unseen, the robot initiates an exploration process by continuously pushing until the thing is located. Following the discovery of the target object, the robot employs coordination of grasp-push actions to successfully grasp it. It is worth noting that this approach assumes prior knowledge and relies on the target object having a specific color. Additionally, the construction of the entire pushing reward function is done manually, potentially necessitating numerous tuning iterations and lacking adaptability to novel scenarios.

Target object-grasping tasks in cluttered environments have been the focus of interest in various research studies [45–50]. In [45], a goal-conditioned hierarchical reinforcement learning approach was proposed, demonstrating high sample efficiency in training a push-to-grasp technique for a particular object amidst clutter. Meanwhile, researchers in [46] focused on extracting a target object from a densely packed environment using quasi-static push and overhand grasp movements. To achieve this, they introduced the visual foresight tree (VFT) method, which identified the shortest sequence of actions. The VFT method combined a deep interactive prediction network (DIPN) to estimate push action outcomes and the Monte Carlo tree search (MCTS) to select the best actions. Despite its effectiveness, the VFT method had some limitations. The computation time was considerably long due to the size of the MCTS tree. In [48], developers introduced a bifunctional network that processes visual observations and produces comprehensive pixel-wise Q value maps for both pushing and grasping primitive actions. This innovation was aimed at augmenting the available data samples within the action space. The system is feasible for practical deployment as the pre-trained model in the simulation achieved a considerably high success rate in the real world without fine-tuning. Investigators in [49] employed Proximal Policy Optimization (PPO) to enable the agent to make predictions about its future actions based on its past experiences and its current state. To facilitate the grasping of occluded objects, a robotic arm equipped with an RGB-D camera was utilized, allowing the agent to observe the scene from various angles. For object tracking using RL, the scene state was encoded using a discretized truncated signed distance field (TSDF) volumetric representation. The study in [50] discusses the difficulty in gripping objects when they are near to one another and there isn't enough room for gripper fingers to do so. They use distinct fully convolutional networks (FCNs) to anticipate the grab point and push direction in their method. Then, using Q-learning, the corresponding action is carried out using the highest Q-value that was chosen. However, rule-based systems are less efficient since the robot could continue pushing after the push operations are completed without changing the robot's workspace, which has an impact on the robot's performance. A common limitation in these studies is the assumption of prior knowledge and reliance on target objects having specific colours, which poses significant constraints in real-world applications.

3.2 Assembly and Disassembly Tasks

Assembly plays a pivotal role in modern manufacturing and production processes. Researchers have delved into robot learning to augment the efficiency and effectiveness of assembly tasks. Several studies on this subject are listed in Table 2.

Table 2. Literature of Deep Reinforcement Learning in Robotic Assembly and Disassembly

Author(s)	Publication year	Task	Method	Observation	Action
Luo et al. [51]	2018	Inserting a peg in a hole	GPS	Force, torque and robot state	Reference to an admittance controller
De Winter et al. [52]	2019	Cranfield assembly benchmark	Q-learning	Labelled state of objects	Hierarchy of sub-procedure
Li et al. [53]	2019	Circuit breaker assembly	DQN	Force and torque	Rotate along three axis
Kristensen et al. [54]	2019	E-waste unscrewing disassembly	Q-learning	Force, joint angle, and position of the end-effector	Seven types of movements
Kim et al. [55]	2020	Inserting a peg in a hole	Imitation learning + DDPG	Force and position	Desired position and velocity
Ota et al. [56]	2019	Computer assembly	TD3	Joint angles and angular velocities	Angular velocities

In [51], researchers were pioneers in applying reinforcement learning to tackle an industrial problem involving deformable objects. Their primary emphasis was on achieving precision. They developed a policy search framework for robotic assembly tasks involving rigid and deformable components. Consequently, they successfully guided a position and velocity-controlled robot with haptic feedback to insert a wooden peg into a non-linear deformable part with a hole. However, it is worth noting that this system did not incorporate a visual perception system, and its performance suffered when the peg was not close to the hole. In subsequent studies, researchers in [52] utilized Q-learning along with a hierarchical task graph for robotic assembly, while investigators in [53] employed DQN and ROS Gazebo simulators to train a KUKA iiwa robot in completing a circuit breaker assembly task. Developers in [54] explored a robot simulation framework for e-waste disassembly using Q-learning, and the study in [55] focused on robot learning for square peg-in-hole assembly, ingeniously combining imitation learning and DDPG. Additionally, investigators in [56] investigated motion trajectory in unknown environments using TD3 with RRT as the reference and validated their approach through a computer assembly task. These studies collectively contribute to the advancement of robotic assembly techniques, showcasing the versatility and efficacy of RL in tackling various challenges across diverse tasks and environments. Their findings pave the way for further innovations and advancements in the field of robotics.

4   Human-Robot Collaboration and Deep Reinforcement Learning

The objective of this review is to offer a comprehensive understanding of the application of reinforcement learning in research related to human-robot collaboration.

4.1   Collaborative Robotics

Human-Robot Collaboration (HRC) is a synergistic approach that combines the strengths of two entities: the power, durability, repeatability, and precision of robots with the intuition, flexibility, problem-solving abilities, and sensory perception of humans. Depending on the scope of application, humans and robots can collaborate in various ways. In this article the meaning of collaboration level is adapted from [57], where the collaboration levels are illustrated in Fig. 3.

Currently, the prevailing approach in industrial production centers around human-robot coexistence, which prioritizes safety. The risk of injuries is notably low because distinct workspaces are designated for humans and robots. In this setup, humans and robots carry out their tasks in separate, dedicated areas.

Within human-robot cooperation, these workspaces are aligned, yet the operations are time-separated and executed sequentially. However, when it comes to collaboration, a different dynamic emerges, as both spatial and temporal separations are diminished. Here, direct interaction between humans and robots becomes common.

In this evolving collaborative landscape, the importance of safety requirements for human-robot cooperation and collaboration is rising. It is crucial to ensure that as humans and robots work more closely together, robust safety measures are in place to safeguard all parties involved.

The combination of robot and human capabilities brings about several significant benefits, including:

- Enhanced Efficiency: Robots often possess faster processing abilities and do not experience fatigue like humans. Humans can leverage the high efficiency of robots to perform tasks quickly and effectively.

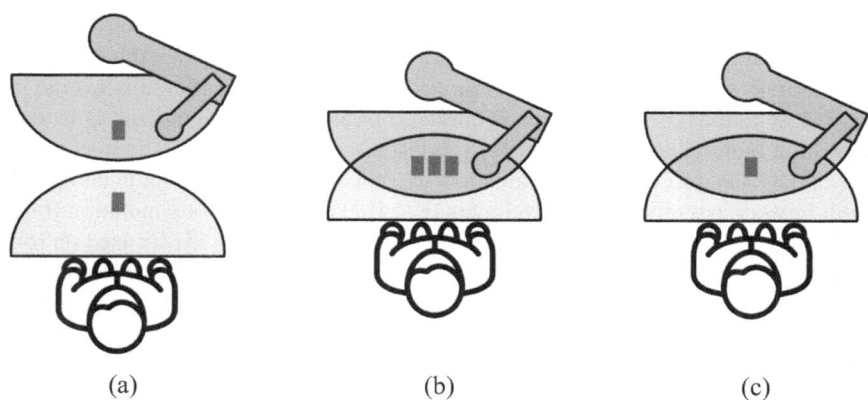

(a)   (b)   (c)

Fig. 3. Collaboration levels adapted for this study: (a) Coexistence; (b) Cooperation; (c) Collaboration.

- Safety in Hazardous Tasks: Robots can be deployed to carry out dangerous, difficult, or repetitive tasks that could pose risks to human safety. When combined with human supervision and control, robots help reduce hazards and enhance safety in hazardous work environments.
- Synergy of Computation and Creativity: The combination of machine computation and human creativity can lead to novel, breakthrough, and innovative solutions in work. Robots can be programmed to handle basic tasks, allowing humans to focus on analyzing and generating more complex solutions.
- Social Interaction with Humans: Robots can be programmed to interact socially with humans, creating a comfortable and easy communication environment in the workplace. This fosters better collaboration and information exchange between humans and robots.

4.2 Deep Reinforcement Learning in Human-Robot Collaboration

In the journey towards smart manufacturing, seamless interaction between humans and industrial robots holds significant importance. Automated robots enhance efficiency and precision, while human involvement introduces a crucial element of adaptability. A prevailing trend in this landscape involves incorporating human factors into interaction strategies to optimize human-robot collaboration. However, the challenge lies in addressing the unpredictability of human behaviors, which can complicate robot task planning and decision-making processes. To tackle this issue, deep reinforcement learning emerges as a vital tool in developing control approaches. Deep RL helps manage the uncertainty introduced by human actions and enhances the robot's ability to adapt to dynamic scenarios (Table 3).

Safety is of paramount importance in manufacturing operations. In human-robot interaction (HRI), one crucial aspect of safety control involves utilizing Deep Reinforcement Learning (DRL) to develop collision avoidance motion plans and navigation control strategies tailored for industrial robot arms. An illustrative example of this approach can be found in the work of researchers highlighted in reference [58]. They addressed safety

Table 3. Literature of Deep Reinforcement Learning in Robotic Grasping

Author(s)	Publication year	Application	Interactive interface	Method	Collaborative level	Observation	Action
Liu et al. [58]	2021	Safe interaction	Vision	DDPG	Cooperation	3D obstacle	Joint angles
Haage et al. [59]	2017	Assembly	Vision	Imitation learning	Coexistence	Human actions visually	Pre-defined optional actions
Zanchettin et al. [60]	2018	Assembly	Vision	Markov Chains	Cooperation	Human actions visually	Accept or not
Wang et al. [61]	2018	Assembly	Audio recognition Vision AR system	Imitation learning	Collaboration	12 assembly task states	Pre-defined movements
Akkaladevi et al. [62]	2019	Assembly	Vision	Q-learning	Collaboration	Human actions visually and object	Sub-procedure
Wu et al. [63]	2020	Object handling	Human reference	Q-learning	Collaboration	Human reference	Object motion
Zhang et al. [64]	2020	Human assistant/Assembly	Vision	RNN	Collaboration	Location of the body joints	Target trajectory of robot
Yu et al. [65]	2021	Assembly	Human reference	DQN	Coexistence	Task chessboard	Sub-procedure
Wang et al. [66]	2021	Path generation	Vision	Imitation learning and Q-learning	Cooperation/ Collaboration	Search parameters	Pre-defined movements
Wang et al. [67]	2021	Hand-over collaboration	Vision	Imitation learning	Collaboration	10 subclasses of hand-over intensions	Accept or not
Zhang et al. [68]	2022	Assembly	Vision	DDPG	Cooperation/ Collaboration	Current state of the robot	Task selection
Deng et al. [69]	2017	Object handling	Vision	Q-learning	Cooperation/ Collaboration	Human actions visually	Joint action
Ghadirzadeh et al. [70]	2016	Object handling	Vision	Q-learning	Cooperation/ Collaboration	Sensory states and actions	Velocities in the end-effector

concerns within industrial human-robot collaboration scenarios by implementing Deep Deterministic Policy Gradients (DDPG). Their innovative approach involved using data from the human arm's position as an observation input. With this information, DDPG generated precise joint angle commands for an ABB IRB1200 robot in real-time, facilitating dynamic trajectory planning to ensure safety and seamless collaboration between humans and the robot.

Additionally, the most talked-about application in production activities is assembly. Deep Reinforcement Learning (DRL) is widely employed as a tailored learning methodology to bolster robot capabilities in providing assembly assistance. A significant proportion of current approaches places their emphasis on aspects like human feedback, human demonstrations, or human-guided training methods. The study in [59] introduced the teach-by-demonstration framework for smartphones, which aims to reduce the time and expertise needed to set up a robotized assembly station. This method allows for changes in roles and tasks, enabling workers to decrease their physical or cognitive workload.

In reference [60], researchers proposed a cutting-edge technique to forecast human activity patterns. By anticipating when a human will request a specific collaborative task, this capacity enables the robot to engage in alternate autonomous tasks proactively. The prediction algorithm at the heart of this strategy is based on higher-order Markov Chains and was extensively tested by experiments carried out in an actual environment. In a broader context, a comprehensive approach is detailed in [61], wherein a three-pronged, integrated strategy encompassing teaching, learning, and operation is adopted. In this approach, humans initially instruct the robot using natural language commands, and subsequently, the robot learns from human assembly demonstrations via a Reinforcement Learning algorithm. Following the teaching and learning phases, the acquired knowledge is actively applied during collaborative assembly tasks to provide valuable assistance.

Academic researchers in [62] employed tabular Q-learning with the reward signal from the human collaborator. In this method, the robot actively scans the work area, understands the assembly procedure, and decides which activities to carry out. The robot system includes 3D sensors to keep an eye on the user and their surroundings and a dynamic graphical user interface (GUI) for user interaction. Additionally, this framework allows for different user types, enabling them to command the robot in various assembly processes. In an analogous situation, as explained in [63], the main goal was to limit item movement to a specific surface. In such cases, people control the robot's end effector in a predetermined area of interest. The robot then modifies its end effector while still being held by the operator, ensuring accurate positioning and orientation alignment for the requested task.

In [64], they utilized sequences of motion frames from the user for experimental validation. They implemented multiple cascaded RNNs, a common approach to using them. The robot observes the human, predicts their next pose, and proactively moves to pick up a screwdriver and hand it to the human based on the prediction. In reference [65], researchers introduced an innovative approach using Deep Q-Networks (DQN) to schedule collaborative tasks between humans and robots efficiently. This method aims to optimize task completion times within a simulated manufacturing environment, specifically in an assembly chessboard scenario. Remarkably, the agent autonomously learns the optimal scheduling policy without any reliance on human intervention or expert knowledge, leveraging a Markov game model to achieve this level of automation. Researchers in [66] proposed a complex method embodying imitation learning, Q-learning, and a simulated annealing algorithm. Gaussian noises were designed in demonstrations to overcome trembling and abrupt changes during a human's demonstrations to avoid jerky regression paths.

Developers in [67] first used multimodal processing to estimate ten subclasses of hand-over intentions and then took them as observation. The robot was trained to decide whether to accept the delivery from the human. Scientists in [68] proposed an approach that utilizes the DDPG method to generate a suitable action sequence for humans and robots during collaborative assembly tasks. The real-time behavior of the agent-human interaction is displayed to the operator. This makes it possible for the operator to complete the assembly work according to the planned assembly behavior, using the globally effective method for the anticipated performance.

In [69], the researchers introduce a hierarchical robot learning approach that consists of two learning hierarchies for HRC tasks. During the mission, the human collaboratively lifts an object with the robot and repeatedly moves it from point A to point B. The robot detects the intention of human and follows its movement, maintaining the thing at a specific orientation throughout the task. In [70], researchers introduced a sensorimotor reinforcement learning framework. This framework is designed to empower robots with the ability to acquire the skills needed for effective collaboration with human partners. The algorithm relies on inputs from vision and force/torque sensors to make informed decisions regarding motor commands. To account for the inherent unpredictability in human actions, a Gaussian process model is employed to model uncertainty. Bayesian optimization is employed to select the most optimal actions at each time step.

5 Open Problems and Research Directions

This section gives a general overview of the difficulties that reinforcement learning currently faces and investigates potential research trajectories for improving teaching strategies in intelligent robotic manufacturing.

A. Sim to Real
Simulation to Reality (Sim2Real) technology aims to transfer knowledge learned in simulators to the real world [71]. This involves setting control tasks, training an agent in a virtual environment provided by a simulator physics engine, and then deploying the acquired policies to control a physical agent in the real world. While training robots in simulators is often straightforward and effective, challenges arise if the simulator fails to accurately represent the complexities of real-world robotic tasks, leading to distribution shifts and failures when deploying the trained robot in the physical environment, despite good performance in the simulated environment. This discrepancy is known as the "sim-to-real gap" in robot learning. Luckily, manufacturing research is actively embracing digital twin technology [72], which holds promise in minimizing the sim-to-real gap in robot learning. The use of digital twins presents an open area of investigation in this context.

B. Multi-agent Reinforcement Learning
When considering robotic manufacturing systems from a higher perspective, the concept of multi-agent multi-task robot learning emerges in the field of artificial intelligence. This method has demonstrated its potential in esports, where multiple agents were trained to function as a virtual army with individual tasks [73]. Extending this approach to robotic manufacturing factories presents an open and challenging opportunity.

A class of techniques known as multi-agent reinforcement learning (MARL) uses reinforcement learning algorithms for individual agents in multi-intelligent systems. In such arrangements, each agent contains fundamental learning, thinking, and planning abilities. By utilizing MARL, an intelligent agent can collaborate with numerous entities that have simpler intelligence to attain complex intelligence, improving system robustness, dependability, and flexibility. Applications for MARL have been found in many fields, including distributed sensing networks, scheduling of transportation, power system optimization, and robot navigation, demonstrating its effectiveness in managing many intelligent systems.

C. Transfer Learning and Generalization

In the realm of manufacturing, environments often exhibit significant variations due to changes in products, production lines, and external conditions. To address this complexity, it is essential to develop reinforcement learning agents capable of transferring knowledge across diverse scenarios and generalizing learned policies to adapt to new settings. However, establishing a knowledge sharing architecture, managing knowledge from different robots, and determining the suitability of specific knowledge pose substantial challenges for researchers in [74]. Additionally, effectively utilizing knowledge gained from previous tasks in new ones remains an open question, especially in the field of smart robotic manufacturing [75]. Overcoming these challenges is crucial for improving the efficiency and adaptability of RL-based systems in manufacturing, unlocking new possibilities for intelligent automation in diverse and dynamic manufacturing environments.

6   Conclusion

Deep Reinforcement Learning (DRL) is a critical and promising technology that holds significant potential across various stages of the smart manufacturing lifecycle. Its adaptability and flexibility make it an appealing solution for enhancing smart manufacturing systems, ushering in a more cognitive and personalized manufacturing approach. To shed light on its essence, this study conducted a systematic literature review of 40 selected works from the past decade, examining the applications of DRL in the engineering product lifecycle. The review emphasizes the challenges faced and proposes future research directions for integrating DRL into smart manufacturing. By providing this comprehensive review, the aim is to inspire further in-depth research and discussions on DRL and its broader implementation in smart manufacturing practices.

Acknowledgement. This research is funded by Vietnam National University-Ho Chi Minh City (VNU-HCM) under grant number: **DS2023-20-02**.

References

1. Nguyen, H., La, H.: Review of deep reinforcement learning for robot manipulation. In: 2019 Third IEEE International Conference on Robotic Computing (IRC), pp. 590–595. IEEE (2019)

2. Ghadirzadeh, A., Chen, X., Yin, W., Yi, Z., Björkman, M., Kragic, D.: Human-centered collaborative robots with deep reinforcement learning. IEEE Robot. Autom. Lett. **6**(2), 566–571 (2020)

3. Fan, L., Zhu, Y., Zhu, J., Liu, Z., Zeng, O., Gupta, A., Fei-Fei, L.: Surreal: Open-source reinforcement learning framework and robot manipulation benchmark. In: Conference on Robot Learning, pp. 767–782. PMLR (2018)

4. Ribeiro, E.G., de Queiroz Mendes, R., Grassi, V., Jr.: Real-time deep learning approach to visual servo control and grasp detection for autonomous robotic manipulation. Robot. Auton. Syst. **139**, 103757 (2021)

5. Ngo, H.Q.T., Bui, T.T.: Application of the image processing technique for powerline robot. In: Phan, C.V., Nguyen, T.D. (eds.) Context-Aware Systems and Applications: 11th EAI International Conference, ICCASA 2022, Vinh Long, Vietnam, October 27-28, 2022, Proceedings, pp. 178–189. Springer Nature Switzerland, Cham (2023). https://doi.org/10.1007/978-3-031-28816-6_14

6. Wang, B., Tao, F., Fang, X., Liu, C., Liu, Y., Freiheit, T.: Smart manufacturing and intelligent manufacturing: a comparative review. Engineering **7**(6), 738–757 (2021)

7. Nguyen, T.T., Nguyen, T.H., Ngo, H.Q.T.: Using real-time operating system to control the recycling waste system in beverage industry for circular economy: mechanical approach. Results Eng. **18**, 101083 (2023)

8. National Research Council: Frontiers in Massive Data Analysis. National Academies Press (2013)

9. Peters, J., Lee, D.D., Kober, J., Nguyen-Tuong, D., Bagnell, J.A., Schaal, S.: Robot learning, pp. 357–398. Springer, Springer handbook of Robotics (2016)

10. Elguea-Aguinaco, Í., Serrano-Muñoz, A., Chrysostomou, D., Inziarte-Hidalgo, I., Bøgh, S., Arana-Arexolaleiba, N.: A review on reinforcement learning for contact-rich robotic manipulation tasks. Robot. Comput. Integr. Manuf. **81**, 102517 (2023)

11. Watkins, C.J., Dayan, P.: Q-learning. Mach. Learn. **8**, 279–292 (1992)

12. Mnih, V., Kavukcuoglu, K., Silver, D., Graves, A., Antonoglou, I., Wierstra, D., Riedmiller, M.: Playing Atari with deep reinforcement learning. arXiv 2013, arXiv:1312.5602

13. Van Hasselt, H., Guez, A., Silver, D.: Deep reinforcement learning with double q-learning. In Proceedings of the AAAI Conference on Artificial Intelligence, the Phoenix Convention Center, vol, 30, Phoenix, AZ, USA (2016)

14. Wang, Z., Schaul, T., Hessel, M., Hasselt, H., Lanctot, M., Freitas, N.: Dueling network architectures for deep reinforcementlearning. In: Proceedings of the International Conference on Machine Learning, pp. 1995–2003, New York City, NY, USA (2016)

15. Sutton, R.S., McAllester, D., Singh, S., Mansour, Y.: Policy gradient methods for reinforcement learning with function approximation. Adv. Neural Inform. Process. Syst. **12** (1999)

16. Schulman, J., Levine, S., Abbeel, P., Jordan, M., Moritz, P.: Trust region policy optimization. In: Proceedings of the International Conference on Machine Learning, Lille, France, pp. 1889–1897 (2015)

17. Schulman, J., Wolski, F., Dhariwal, P., Radford, A., Klimov, O. In: Proximal policy optimization algorithms. arXiv 2017 arXiv:1707.06347

18. Konda, V.R., Tsitsiklis, J.N.: Actor-critic algorithms. In: Proceedings of the Advances in Neural Information Processing Systems, Denver, CO, USA, 28–30, pp. 1008–1014 (2000)

19. Mnih, V. et al: Asynchronous methods for deep reinforcement learning. In: Proceedings of the International Conference on Machine Learning, New York City, NY, USA 19–24 1928 1937 (2016)

20. Lillicrap, T.P., et al.: Continuous control with deep reinforcement learning. arXiv 2015 arXiv: 1509.02971

21. Silver, D., Lever, G., Heess, N., Degris, T., Wierstra, D., Riedmiller, M.: Deterministic policy gradient algorithms. In International conference on machine learning (pp. 387–395). PMLR (2014)
22. Fujimoto, S., Hoof, H., Meger, D.: Addressing function approximation error in actor-critic methods. In: Proceedings of theInternational Conference on Machine Learning, Stockholm, Sweden, 10–15, pp. 1587–1596 (2018)
23. Haarnoja, T., et al. Soft actor-critic algorithms and applications. arXiv 2018 arXiv:1812.05905
24. Ngo, H.Q.T.: Design of automated system for online inspection using the convolutional neural network (CNN) technique in the image processing approach. Results Eng., 101346 (2023)
25. Phan, L.A.D., Ngo, H.Q.T.: Application of the artificial intelligence technique to recognize and analyze from the image data. In: Deep Learning and Other Soft Computing Techniques: Biomedical and Related Applications (pp. 77–89). Springer Nature Switzerland, Cham. https://doi.org/10.1007/978-3-031-29447-1_8
26. Sperduti, A., Starita, A. Supervised neural networks for the classification of structures. IEEE Trans. Neural Netw. **8**, 714–735 (1997)
27. Morris, C., et al.: Weisfeiler and Leman go neural: higher-order graph neural networks. In: Proceedings of the Association for the Advancement of Artificial Intelligence (AAAI) Conference and Artificial Intelligence, Honolulu, HI, USA, vol. 33, pp. 4602–4609 (2019)
28. Hamilton, W.L., Ying, R., Leskovec, J.: Inductive representation learning on large graphs. In: Proceedings of the 31st International Conference on Neural Information Processing Systems, Long Beach, CA, USA, pp. 1025–1035 (2017)
29. Veličković, P., Cucurull, G., Casanova, A., Romero, A., Lio, P., Bengio, Y. Graph attention networks. arXiv 2017 arXiv:1710.10903
30. Beltran-Hernandez, C.C., Petit, D., Ramirez-Alpizar, I.G., Harada, K.: Learning to grasp with primitive shaped object policies. In: 2019 IEEE/SICE International Symposium on System Integration (SII) (pp. 468–473). IEEE (2019)
31. Mousavian, A., Eppner, C., Fox, D.: 6-dof graspnet: variational grasp generation for object manipulation. In: Proceedings of the IEEE/CVF International Conference on Computer Vision, pp. 2901–2910 (2019)
32. Berscheid, L., Meißner, P., Kröger, T.: Self-supervised learning for precise pick-and-place without object model. IEEE Robot. Automation Lett. **5**(3), 4828–4835 (2020)
33. Zeng, A., Song, S., Lee, J., Rodriguez, A., Funkhouser, T.: Tossingbot: learning to throw arbitrary objects with residual physics. IEEE Trans. Rob. **36**(4), 1307–1319 (2020)
34. Shao, Q., et al.: Suction grasp region prediction using self-supervised learning for object picking in dense clutter. In 2019 IEEE 5th International Conference on Mechatronics System and Robots (ICMSR), pp. 7–12. IEEE (2019)
35. Zakka, K., Zeng, A., Lee, J., Song, S.: Form2fit: learning shape priors for generalizable assembly from disassembly. In: 2020 IEEE International Conference on Robotics and Automation (ICRA), pp. 9404–9410. IEEE (2020)
36. Cao, H.G., Zeng, W., Wu, I.C.: Reinforcement learning for picking cluttered general objects with dense object descriptors. In: 2022 International Conference on Robotics and Automation (ICRA), pp. 6358–6364. IEEE (2022)
37. Zeng, A., et al.: Robotic pick-and-place of novel objects in clutter with multi-affordance grasping and cross-domain image matching. Int. J. Robot. Res. **41**(7), 690–705 (2022)
38. Zeng, A., Song, S., Welker, S., Lee, J., Rodriguez, A., Funkhouser, T.: Learning synergies between pushing and grasping with self-supervised deep reinforcement learning. In: 2018 IEEE/RSJ International Conference on Intelligent Robots and Systems (IROS) (pp. 4238–4245). IEEE (2018)
39. Ren, D., Ren, X., Wang, X., Digumarti, S.T., Shi, G.: Fast-Learning Grasping and Pre-Grasping via Clutter Quantization and Q-map Masking. In: 2021 IEEE/RSJ International Conference on Intelligent Robots and Systems (IROS), pp. 3611–3618. IEEE (2021)

40. Tang, B., Corsaro, M., Konidaris, G., Nikolaidis, S., Tellex, S.: Learning collaborative pushing and grasping policies in dense clutter. In 2021 IEEE International Conference on Robotics and Automation (ICRA), pp. 6177–6184. IEEE (2021)

41. Zhang, H., et al.: Reinforcement learning based pushing and grasping objects from ungraspable poses. arXiv preprint arXiv:2302.13328 (2023)

42. Berscheid, L., Meißner, P., Kröger, T.: Robot learning of shifting objects for grasping in cluttered environments. In 2019 IEEE/RSJ International Conference on Intelligent Robots and Systems (IROS), pp. 612–618. IEEE (2019)

43. Hundt, A., et al.: Good robot!: efficient reinforcement learning for multi-step visual tasks with sim to real transfer. IEEE Robot. Autom. Letters **5**(4), 6724–6731 (2020)

44. Yang, Y., Liang, H., Choi, C.: A deep learning approach to grasping the invisible. IEEE Robot. Autom. Lett. **5**(2), 2232–2239 (2020)

45. Xu, K., Yu, H., Lai, Q., Wang, Y., Xiong, R.: Efficient learning of goal-oriented push-grasping synergy in clutter. IEEE Robot. Autom. Lett. **6**(4), 6337–6344 (2021)

46. Huang, B., Han, S.D., Yu, J., Boularias, A.: Visual foresight trees for object retrieval from clutter with nonprehensile rearrangement. IEEE Robot. Autom. Lett. **7**(1), 231–238 (2021)

47. Chebotar, Y., et al.: Actionable models: unsupervised offline reinforcement learning of robotic skills. arXiv preprint arXiv:2104.07749 (2021)

48. Ren, D., Wu, S., Wang, X., Peng, Y., Ren, X.: Learning bifunctional push-grasping synergistic strategy for goal-agnostic and goal-oriented tasks. arXiv preprint arXiv:2212.01763 (2022)

49. Novkovic, T., Pautrat, R., Furrer, F., Breyer, M., Siegwart, R., Nieto, J.: Object finding in cluttered scenes using interactive perception. In 2020 IEEE International Conference on Robotics and Automation (ICRA), pp. 8338–8344. IEEE (2020)

50. Chen, Y., Ju, Z., Yang, C.: Combining reinforcement learning and rule-based method to manipulate objects in clutter. In: 2020 International Joint Conference on Neural Networks (IJCNN), pp. 1–6. IEEE (2020)

51. Luo, J., Solowjow, E., Wen, C., Ojea, J.A., Agogino, A.M.: Deep reinforcement learning for robotic assembly of mixed deformable and rigid objects. In: 2018 IEEE/RSJ International Conference on Intelligent Robots and Systems (IROS), pp. 2062–2069. IEEE (2018)

52. De Winter, J., De Beir, A., El Makrini, I., Van de Perre, G., Nowé, A., Vanderborght, B.: Accelerating interactive reinforcement learning by human advice for an assembly task by a cobot. Robotics **8**(4), 104 (2019)

53. Li, F., Jiang, Q., Zhang, S., Wei, M., Song, R.: Robot skill acquisition in assembly process using deep reinforcement learning. Neurocomputing **345**, 92–102 (2019)

54. Kristensen, C.B., Sørensen, F.A., Nielsen, H.B., Andersen, M.S., Bendtsen, S.P., Bøgh, S.: Towards a robot simulation framework for e-waste disassembly using reinforcement learning. Procedia Manuf. **38**, 225–232 (2019)

55. Kim, Y.L., Ahn, K.H., Song, J.B.: Reinforcement learning based on movement primitives for contact tasks. Robot. Comput. Integrated Manuf. **62**, 101863 (2020)

56. Ota, K., Jha, D. K., Oiki, T., Miura, M., Nammoto, T., Nikovski, D., & Mariyama, T. (2019, November). Trajectory optimization for unknown constrained systems using reinforcement learning. In 2019 IEEE/RSJ international conference on intelligent robots and systems (IROS) (pp. 3487–3494). IEEE

57. Müller, R., Vette, M., Geenen, A.: Skill-based dynamic task allocation in human-robot-cooperation with the example of welding application. Procedia Manufacturing **11**, 13–21 (2017)

58. Liu, Q., Liu, Z., Xiong, B., Xu, W., Liu, Y.: Deep reinforcement learning-based safe interaction for industrial human-robot collaboration using intrinsic reward function. Adv. Eng. Inform. **49**, 101360 (2021)

59. Haage, M., et al.: Teaching assembly by demonstration using advanced human robot interaction and a knowledge integration framework. Procedia Manufacturing **11**, 164–173 (2017)
60. Zanchettin, A.M., Casalino, A., Piroddi, L., Rocco, P.: Prediction of human activity patterns for human–robot collaborative assembly tasks. IEEE Trans. Industr. Inf. **15**(7), 3934–3942 (2018)
61. Wang, W., Li, R., Chen, Y., Diekel, Z.M., Jia, Y.: Facilitating human–robot collaborative tasks by teaching-learning-collaboration from human demonstrations. IEEE Trans. Autom. Sci. Eng. **16**(2), 640–653 (2018)
62. Akkaladevi, S.C., Plasch, M., Pichler, A., Ikeda, M.: Towards reinforcement based learning of an assembly process for human robot collaboration. Procedia Manufacturing **38**, 1491–1498 (2019)
63. Wu, M., He, Y., Liu, S.: Adaptive impedance control based on reinforcement learning in a human-robot collaboration task with human reference estimation. Int J Mech Control **21**(1), 21–31 (2020)
64. Zhang, J., Liu, H., Chang, Q., Wang, L., Gao, R.X.: Recurrent neural network for motion trajectory prediction in human-robot collaborative assembly. CIRP Ann. **69**(1), 9–12 (2020)
65. Yu, T., Huang, J., Chang, Q.: Optimizing task scheduling in human-robot collaboration with deep multi-agent reinforcement learning. J. Manuf. Syst. **60**, 487–499 (2021)
66. Wang, Y.Q., Hu, Y.D., El Zaatari, S., Li, W.D., Zhou, Y.: Optimised learning from demonstrations for collaborative robots. Robotics and Computer-Integrated Manufacturing **71**, 102169 (2021)
67. Wang, W., Li, R., Chen, Y., Sun, Y., Jia, Y.: Predicting human intentions in human–robot handover tasks through multimodal learning. IEEE Trans. Autom. Sci. Eng. **19**(3), 2339–2353 (2021)
68. Zhang, R., Lv, Q., Li, J., Bao, J., Liu, T., Liu, S.: A reinforcement learning method for human-robot collaboration in assembly tasks. Robotics and Computer-Integrated Manufacturing **73**, 102227 (2022)
69. Deng, Z., Mi, J., Han, D., Huang, R., Xiong, X., Zhang, J.: Hierarchical robot learning for physical collaboration between humans and robots. In 2017 IEEE international conference on robotics and biomimetics (robio), pp. 750–755. IEEE (2017)
70. Ghadirzadeh, A., Bütepage, J., Maki, A., Kragic, D., Björkman, M.: A sensorimotor reinforcement learning framework for physical human-robot interaction. In: 2016 IEEE/RSJ International Conference on Intelligent Robots and Systems (IROS), pp. 2682–2688. IEEE (2016)
71. Zhao, W., Queralta, J.P., Westerlund, T.: Sim-to-real transfer in deep reinforcement learning for robotics: a survey. In 2020 IEEE symposium series on computational intelligence (SSCI), pp. 737–744. IEEE (2020)
72. Tao, F., Qi, Q.: Make more digital twins. Nature **573**(7775), 490–491 (2019)
73. Vinyals, O., et al.: Grandmaster level in StarCraft II using multi-agent reinforcement learning. Nature **575**(7782), 350–354 (2019)
74. De Bruin, T., Kober, J., Tuyls, K., Babuska, R.: Experience selection in deep reinforcement learning for control. J. Mach. Learn. Res. **19** (2018).
75. Zhao, T. Z., Luo, J., Sushkov, O., Pevceviciute, R., Heess, N., Scholz, J., Levine, S.: Offline meta-reinforcement learning for industrial insertion. In: 2022 International Conference on Robotics and Automation (ICRA) (pp. 6386–6393). IEEE (2022)

User-Based Collaborative Filtering Multi-criteria Recommender System Based on Interaction Between Criteria, Criteria Set with Choquet Integral

Tri Minh Huynh[1], Vu The Tran[2], and Hiep Xuan Huynh[3(✉)]

[1] Kien Giang University, Kien Giang, Vietnam
[2] University of Science and Technology, Da Nang University, Da Nang, Vietnam
[3] Can Tho University, Can Tho, Vietnam
hxhiep@ctu.edu.vn

Abstract. The exploitation of knowledge in stored data is one of the current research trends. The increasing of the need of users about searching information, the consulting system is special attented by researchers. Many decision-making solutions for multi-criteria recommender model have been proposed. However, with the intrinsic of the data, the latent values of the interaction relationship, the dominance between the criteria always changes the results of decision-making to advise users. When we take enought these values, decision-making becomes more efficient. In this paper, we propose a new approach to building a decision model for a user-based multi-criteria filtering system with Choquet integration. The operation is also based on the capacity function of a criterion, a set of criteria. The effect of decision making is the degree of interaction between the criteria in the data. This model is also based on traditional techniques and integrates some our new methods. We tested and evaluated the proposed model on the multirecsys tool we built. The standard datasets is used to test. We compare results with some the same existing models. Through experimentation, we saw that the proposed model is quite effective and reliable. It can be applied well in many appropriate systems, contributing to improving the deficiencies and the limitations of the current recommendation.

Keywords: Recommender · multi-criteria · user-based · interaction · Choquet

1 Introduction

Recommender systems [1] are growing and having an important role in our lives. It helps users find information quickly according to their references. There are many different consulting models that service different requirements. Depending on the characteristics of each system, we can apply the appropriate model. Many decision-making solutions can be based on historical factors, habits, popularity, correlation… in stored data. Today, many decision-making models have not attended much to the hidden values in data.

© ICST Institute for Computer Sciences, Social Informatics and Telecommunications Engineering 2024
Published by Springer Nature Switzerland AG 2024. All Rights Reserved
P. Cong Vinh and N. Thanh Tung (Eds.): ICCASA 2023, LNICST 579, pp. 43–53, 2024.
https://doi.org/10.1007/978-3-031-58878-5_3

In other hand, searching knowledge that exists and is reflected in the data to make better recommendations for users. One of the solutions for that goal is applying Choquet integration to calculate the interaction values in the data. The multi-criteria recommender system [2, 3, 5, 17–20] is the most research choice today because user's preferences are always diverse and data is increased more, so it is necessary to consider on many criteria to make a more effective decision. Many operations have been applied to decision making for multi-criteria recommender model [2, 16, 19] but has'n been considered to the interaction between criteria and sets of criteria each other.

In this article, we propose a new approach to build user-based collaborative filtering multi-criteria recommender system based on interaction between criteria, criteria set with Choquet integration. A capacity function is applied to perform fuzzy measurements in the applicable data for calculations. The criteria and set of criteria are calculated about the level of interaction in the decision-making operation but there is a limit on the number of criteria sets. We build the model only make with at most 3 criteria because based on the ability of the empirical tool, the model could not calculate all sets of criteria in the system. Three standard data sets were applied to experiment the model are: MovieLense, MSWeb and Jester5k, this data set has different nature.

2 Multi-criteria Decision

2.1 Multi-criteria Decision Model

We build the model with matrix $M(x \times y)$ consists of x rows $r_1, r_2, \ldots, r_x$ and y columns $c_1, c_2, \ldots, c_y$. Each row of $r_i(i : 1..x)$ with each column of $c_q(j : 1..y)$ is determined the value $r_{ij} = (r_i, c_j)$ as Table 1. A criterion is defined be a row or a column. Call sets: $R = \{r_1, r_2, \ldots, r_x\}$ and $C = \{c_1, c_2, \ldots, c_y\}$, define a function $\hat{r} : \mathcal{F}(RxC) \to \mathbb{R}$ determines values $\hat{r}_i$ on M from set R and C.

Table 1. Multi-criteria decision model

row/column	r_1	r_2	...	r_x
c_1	5	1	...	3
c_2	3	3	...	5
c_3	3	4	...	1
...	...	...	...	...
c_y	2	3	...	4
$\hat{r}$	$\hat{r}_1 = 63$	$\hat{r}_2 = 38$	...	$\hat{r}_x = 45$

Example: $\hat{r_1} = (r_1xC)$ is determined to be the sum of values $\{r_{11}, r_{12}, \ldots, r_{1y}\}$

$$\hat{r}_1 = \mathcal{F}(r_1 \times C) = \sum_{j=1}^{x} r_{1j} = \mathcal{F}(\{5, 3, 3, \ldots, 2\}) = 63$$

R and C is called multi-criteria sets, $\hat{r}$ is called a multi-criteria decision function and this model is called the multi-criteria decision model [1][6][7].

2.2 Model Multiple-Criteria Model with Some Operations Are Used for Decision Making: Arithmetic Mean (AM), Geometric Mean (GM), Harmonic Mean (HM) and Ordered Weighted Averaging Operator (OWA)

Model have criteria: 3 columns and 4 rows, p: 1..3, q:1..4

AM: $\hat{r}_p\left(r_{p1}, r_{p2}, r_{p3}, r_{p4}\right) = \frac{1}{4}\sum_{q=1}^{4} r_{pq}$

GM: $\hat{r}_p\left(r_{p1}, r_{p2}, r_{p3}, r_{p4}\right) = \sqrt[4]{r_{p1} * r_{p2} * r_{p3} * r_{p4}}$

HM: $\hat{r}_p\left(r_{p1}, r_{p2}, r_{p3}, r_{p4}\right) = \frac{4}{\sum_{i=1}^{4}\left(\frac{1}{r_{pi}}\right)}$

OWA: $\hat{r}_p\left(r_{p1}, r_{p2}, r_{p3}, r_{p4}\right) = 1/4\sum_{j=1}^{4} w_j * r_{pj}$ *(index on w and r)*.

Table 2. Some Operations Are Used For Decision Making

row/column	r_1	r_2	r_3	w
c_1	5	1	3	1.5
c_2	3	3	5	2.4
c_3	3	4	2	1.3
c_4	2	3	4	4.2
$\hat{r}_{AM}$	3.25	2.75	3.5	
$\hat{r}_{GM}$	3.08	2.45	3.31	
$\hat{r}_{HM}$	2.92	2.08	3.12	
$\hat{r}_{OWA}$	8.82	7.45	9.43	

Decision-making is based on the interaction between criteria and sets of criteria.

2.3 Capacity Function

With set of values $R_p = \{r_{p1}, r_{p2}, r_{p3}, r_{p4}\}$, a capacity function μ 14 on $\hat{r}_p$ is function $\mu : \Re(R_p) \rightarrow [0, 1]$, with $\mu(\emptyset) = 0$, $\mu(R_p) = 1$. [8–13] With A and B are sets of criteria. $A, B \subseteq R_p$ and $A \subseteq B \Rightarrow \mu(A) \leq \mu(B)$. On set R_p, define a vector P with weights, $Q \subseteq R_p$, when μ defined as follow: $\mu(Q) = \sum_{a\in Q} P(a)$,

$$\sum_{i\in R_p} P(i) = 1, i \in R_p \tag{1}$$

The value of $\mu(Q)$ depend on the criteria in Q. C_1, C_2 are two criteria in Q. The value of $\mu(C_1, C_2)$ can get the value as follow:

or $\mu(C_1, C_2) = \mu(C_1) + \mu(C_2)$, or $\mu(C_1, C_2) > \mu(C_1) + \mu(C_2)$, or $\mu(C_1, C_2) < \mu(C_1) + \mu(C_2)$.

Example with three criteria: C_1, C_2, C_3 and $\mu(C_1) = 0.28$, $\mu(C_2) = 0.32$, $\mu(C_3) = 0.47$. To subsets: $\mu(C_1, C_2) = 0.45$, $\mu(C_1, C_3) = 0.89$, $\mu(C_2, C_3) = 0.65$, $\mu(C_1, C_2, C_3) = 1$

2.4 Choquet Fuzzy Integral

With capacity function μ, the Choquet fuzzy integeal [10] C based on μ of $R_p = \{r_{p1}, r_{p2}, r_{p3}, r_{p4}, r_{pm}\}$ is defined by function $\Gamma : R_p \rightarrow \Re^+$, $\mu(A)$ is fuzzy measure on subset A, $\mu(R_p)$ is fuzzy measure of R_p. Choquet integral is one of the operations that show clearly about the interactive relationship between criteria:

$$C(\Gamma) = \sum_{q=1}^{m}(\Gamma(r_{pq}) - \Gamma(r_{p(q-1)}))\mu(A_{pq}) \tag{2}$$

where $A_{pq} = \{r_{pq}, r_{p(q+1)}, \ldots, r_{pm}\} \in R_p$ is set of k criteria, with $k = m - p + 1$: satisfying the conditions: $0 \geq \Gamma(r_{p1}) \geq \cdots \geq \Gamma(r_{pm}) \geq 1 and \Gamma(r_{p0}) = 0$

Example 1: a set of criteria: $\{A, B, C\}$ with the values: $x_A \geq x_B \geq x_C$, the item's aggregated score is calculated based on the Choquet fuzzy integral:

$$C(x_A, x_B, x_C) = x_C * \mu(A, B, C) + (x_B - x_C) * \mu(A, B) + (x_A - x_B) * \mu(A)$$

With $x_A = 5, x_B = 3, x_C = 1, \mu(0) = 0, \mu(A, B, C) = 1, \mu(A) = 0.4, \mu(B) = 0.8, \mu(C) = 0.7, \mu(A, B) = 0.9$. Then $C(x_A, x_B, x_C) = 3.6$.

The capacity function of the criteria set depends on the capacity function of each criterion and the interaction value of criteria. With set of two criteria:

$$\mu(r_{pi}, r_{pj}) = \mu(r_{pi}) + \mu(r_{pj}) + I(r_{pi}, r_{pj})$$

where $I(r_{pi}, r_{pj})$ is call: the interaction value between r_{pi} and r_{pj}, $I(r_{pi}, r_{pj})$ in [-1,1]. When two criteria r_{pi}, r_{pj} are in larger set $B = A \cup \{r_{pi}, r_{pj}\}$ with m criteria:

$$I(r_{pi}, r_{pj}) = \sum_{A \in I \setminus \{r_{pi}, r_{pj}\}} \frac{(m - |A| - 2)|A|!}{m!}[\mu(A \cup \{r_{pi}, r_{pj}\}) - (\mu(A \cup \{r_{pi}\}) + \mu(A \cup \{r_{pj}\})) + \mu(A)]$$

Thus, with each set of criteria $A \in \{(r_{p1}, r_{p2}, \ldots, r_{pm})\}$, the capacity function of A can determine as follow:

$$\mu(A) = \sum_{r_{pi} \in A} \mu(r_{pi}) + I, with I > 0 or I < 0, \tag{3}$$

$if \mu(A) > 1 then \mu(A) = 1$, I is sum of the interaction value of subsets in A.

3 Proposed Model

3.1 Rating Matrix

The model uses the above rating matrix to represent a list of users who rate data items through rows and columns. Items that are not rated will have a value of "?". Here, u_a is the consulted user (Table 3).

Table 3. Data model with Rating matrix

	i_1	i_2	...	i_x	...	i_y	...	i_n
u_1	?	4	...	3	...	?	...	3
u_2	3	4	...	2	...	5	...	4
...	...	...	...	...	...	...	...	...
u_m	5	3	...	4	...	5	...	?
u_a	?	3	...	?	...	?	...	4
$\hat{r}$	?	-	...	?	...	?	...	-

3.2 Similarity

The model uers k nearest neighbors (kNN) [15] to value the similarity (or distance) between u_q (q:1..m) ua is accorded by measures Pearson. The Pearson measure [23] between two items are u_x and u_y is defined:

$$sim_{pearson}\left(u_x, u_y\right) = \frac{\sum_{i \in I_{u_x.u_y}}\left(r_{u_x i} - \bar{r}_{u_x}\right)\left(r_{u_y i} - \bar{r}_{u_y}\right)}{\sqrt{\sum_{i \in I_{u_x.u_y}}\left(r_{u_x i} - \bar{r}_{u_x}\right)^2}\sqrt{\sum_{i \in I_{u_x.u_y}}\left(r_{u_y i} - \bar{r}_{u_y}\right)^2}} \tag{4}$$

I_u is the set of data items evaluated by u_x, $\bar{r}_{u_x}$ is the average rating evaluation of u_x on all data items, $\bar{r}_{u_y}$ is the average rating evaluation of u_y on all data items. Then, the distance between two users is (1-r).

3.3 Determining the Capacity Function of Criteria, a Set of Criteria

(a) Determining $\mu(u_q)$ of each user u_q, $q : 1..m$ in system is a potential weight of each u_q, it is the ratio between the number of the user's rating values (# "?") for item of m users.

$$\mu(u_q) = \frac{count\left(rating\left(u_q, i_p\right)\#"?"\right)}{m} \tag{5}$$

(b) Determined $\mu(A)$ of each subset A, $A \subseteq U$, U is a set of all users, with steps:

- First, $\mu'(u_i)$, $u_i \in A = [\mu(u_i)/ \sum_{j=1}^{s} \mu\left(u_j\right)$, with s is number of u_i in A.
- Determined

$$\mu(A) = sum\left(\mu'(u_i)\right), u_i \in A \tag{6}$$

This value $\mu(A)$ based on μ' and it responds with formula 3. However, the level of interaction in the system is not high.

3.4 Recommendation Model

With a data table as Table 2, a user u_a, is recommended user. A column is used for determining the capacity functions $\mu(u_q)$, $q : 1..m$. At each rating value of $u_a for items =$ "?", we determined the values $\hat{r}_p, p : 1..n$ (Fig. 1, 2, 3, 4, 5, 6, 7).

Table 4. Proposed model.

	i_1	i_2	...	i_8	...	i_{52}	...	i_n	μ
u_1	?	1	...	5	...	?	...	1	
u_2	2	3	...	4	...	2	...	2	0.34
...	...	...	...	...	...	...	...	...	...
u_x	5	2	...	2	...	4	...	5	0.25
...		...		...		...	...		...
U_y	3	4	...	0	...	3	...	1	0.12
...	...	...	...	...	...	...	...	...	...
u_m	3	?	...	4	...	4	...	?	
u_a	3	?	...	?	...	?	...	3	
$\hat{r}$	-	3.04	...	2.16	...	1.2	...	-	

3.5 Identify Results of Recommender System

(1) First, determine the similarity between the user u_a and each user in the data (formula 4). The results are as Table 4,.
(2) Next, determining the capacity function of each user μ (formula 5) in the system. we take the similarity values of kNN (k highest values) to calculate μ by at each $i_p, : 1..n$ with each $u_k, : 1..kNN$.
(3) Next, we calculate the values $\hat{r}_p$ at r_{pk}#?, k : 1..kNN (formula 2,6) at the values of $u_a = ?$. $\hat{r}_p$ values are ranked in descending order, selecting hightest values: i_2 and i_8; $\hat{r}_2 = 3.04 and \hat{r}_8 = 2.16$.

3.6 Evaluation Recommendations

Method used to evaluate model is the Receiver Operating Characteristic (ROC) [4, 16, 17]. Evaluation for two systems can compare the size of the area under the ROC-curve, where a bigger area indicates better performance. Four values contain the true-false positives/negatives, as follows: True Positives (TP): and False Positives (FP). False Negatives (FN) and True Negatives (TN). True Positive Rate TPR = TP/(TP + FN). False Positive Rate FPR = FP/(FP + TN).

$$Precision(L) = \frac{1}{|U|} \sum_{u \in U} |L(u) \cap T(u)| / |L(u)|$$

$$Recall(L) = \frac{1}{|U|} \sum_{u \in U} |L(u) \cap T(u)| / |T(u)|$$

U is set of users. $L(u_a)$ get items that u_a want to chose. $I = I_{test} \cup I_{train}$, $T(u) \subset I_u \cap I_{test}$. I_u is item set that u_a has chosen

4   Experiment

4.1   Data Sets

Experimentation use three datasets: MovieLens100K, MSWeb, Jester5k are integrated in recommenderlab [14].

4.2   Tools

The model was experimented by multirecsys tool which we built, developed and installed applications on R [www.r-project.org].

4.3   Scenario 1: Experiment the Model and Compare It with Some Existing Models

In this scenario, we test the proposed model UBCF_Choquet on three datasets: Movie-lens100K (non-binary dataset), MSWeb (binary dataset) that are two too sparse datasets and Jester5k which is too thick dataset. Using them shows the recommendation results and the existing models (UBCF, IBCF, Random, SVD, Pupular) [2][5]. Each model, we chose five items to recommend for tow users and show ROC-curve and Precision/Recall of each model on three datasets with kNN = 7, the result of the model as follow: On Movilense:

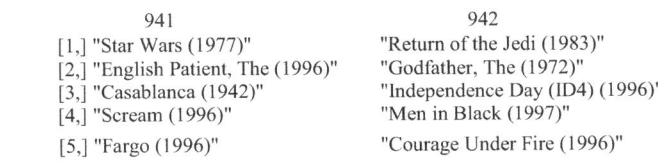

	941	942
[1,]	"Star Wars (1977)"	"Return of the Jedi (1983)"
[2,]	"English Patient, The (1996)"	"Godfather, The (1972)"
[3,]	"Casablanca (1942)"	"Independence Day (ID4) (1996)"
[4,]	"Scream (1996)"	"Men in Black (1997)"
[5,]	"Fargo (1996)"	"Courage Under Fire (1996)"

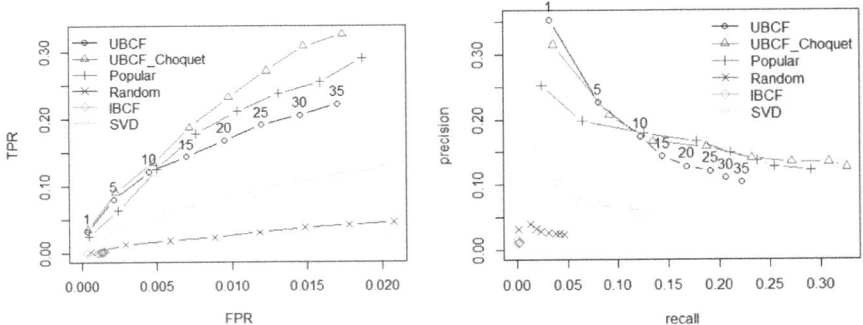

Fig. 1. Roc-curve and Precision/Recall with kNN = 7 on Movielense

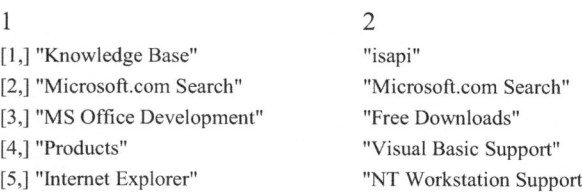

1	2
[1,] "Knowledge Base"	"isapi"
[2,] "Microsoft.com Search"	"Microsoft.com Search"
[3,] "MS Office Development"	"Free Downloads"
[4,] "Products"	"Visual Basic Support"
[5,] "Internet Explorer"	"NT Workstation Support"

On MSWeb:

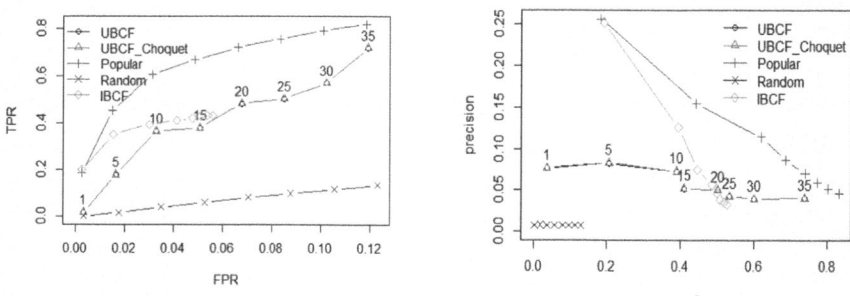

Fig. 2. Roc-curve and Precision/Recall with kNN = 7 on MSWeb.

$u3270 : "j27" "j69" "j38" "j47" "j93"
$u15348: "j76" "j92" "j97" "j93" "j74"

On Jester5k"

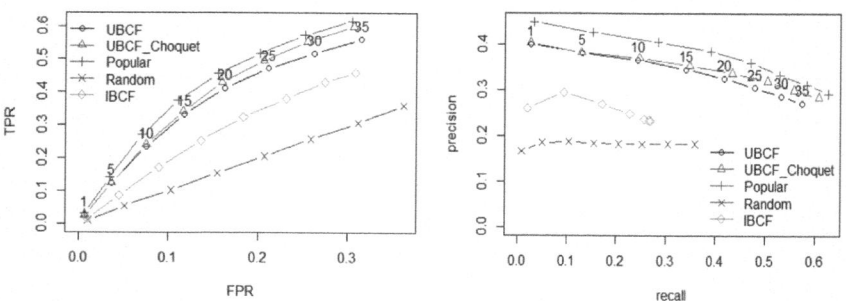

Fig. 3. Roc-curve and Precision/Recall with kNN = 7 on Jester5k

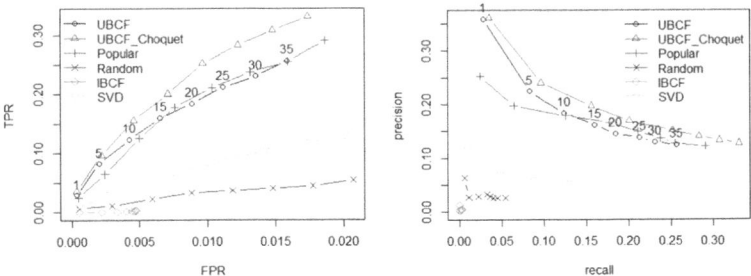

Fig. 4. Roc-curve and Precision/Recall with kNN = 15

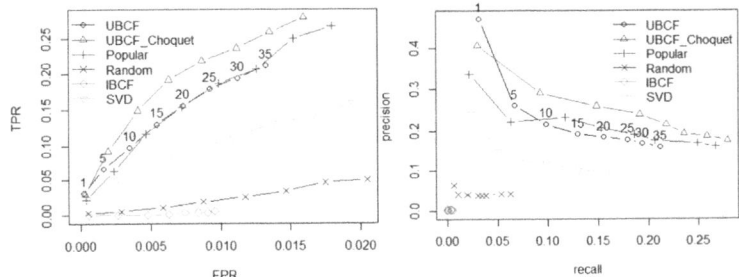

Fig. 5. Roc-curve and Precision/Recall with kNN = 25

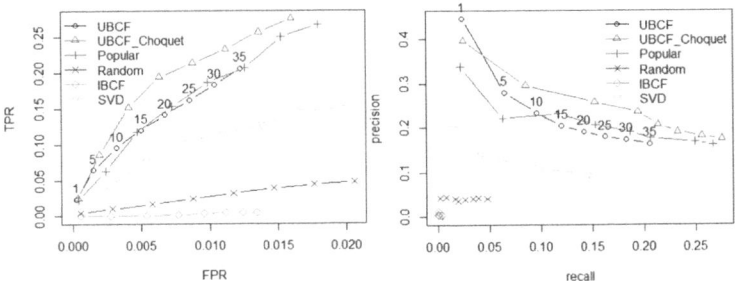

Fig. 6. Roc-curve and Precision/Recall with kNN = 35

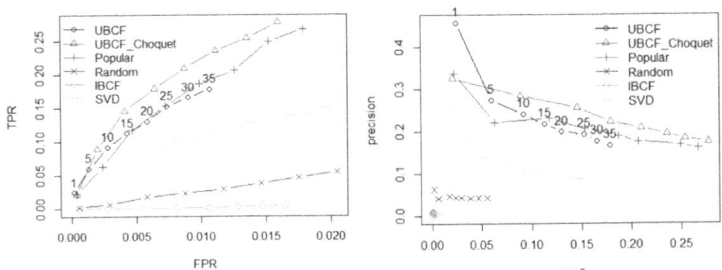

Fig. 7. Roc-curve and Precision/Recall with kNN = 45

4.4 Scenario 2: With Different kNN Values.

4.5 Discussion

The exploitation of the interaction relationships of criteria, sets of criteria for decision making is essential. However, to make effective decisions, we need to fully account for these interactive relationships. In the implementation process, we encountered certain difficulties when the complex exploded without a better solution. We will try to continue to improve in the future. We found that, with three empirical datasets, the proposed model gave rather good results. Especially with sparse or "long tail" datasets, the proposed model gets better consultation results.

5 Conclusions

There are many decision-making solutions for the consulting system. Models often use average operations to calculate values. In this model, we use the Choquet fuzzy integral to calculate for decision making. Data always contains interactive relationships. The average operation ignores these interaction values, so it is not possible to fully evaluate the resonance of the criteria. However, we have only calculated the resonance value of a set of 2 criteria, 3 criteria. Due to the current conditions, we cannot calculate sets that have more criteria. Therefore, the model has not promoted the effect of operations well yet. However, the proposed model is always responsive well, can be applied in many different datasets and contributes to improving the deficiencies of the current consulting system[1].

Acknowledgment. We built this model based on traditional consulting models but there are some changes in the solution. We also exploit and develop on the recommender lab tool package that researchers have developed.

References

1. Michael, D.E., John, T.R., Joseph, A.K.: Collaborative filtering recommender systems. Hum. Comput. Interact. **4**, pp. 291–324. (2011). https://doi.org/10.1561/1100000009
2. Tri, H.M., Vu, T.T., Hung, H.H., Hiep, X.H.: Decision making operations: arithmetic mean, geometric mean, harmonic mean in user-based collaborative filtering of multi-criteria recommender systems. In: The 6th national scientific conference on information technology and applications in various fields, Da Nang, VietNam (2017).
3. Badrul, S., George, K., Joseph, K., John, R.: Item-based Collaborative Filtering Recommendation Algorithms. Appears in WWW10, May 1–5, Hong Kong (2001)
4. Michael, H.: Recommenderlab: A Framework for Developing and Testing Recommendation Algorithms, R package version 0.2–2 (2007). http://lyle.smu.edu/IDA/recommenderlab/
5. Tri, H.M., Vu, T.T., Hung, H.H., Hiep, X.H.: Collaborative filtering recommender system base on the interaction multi-criteria decision with ordered weighted averaging operator. In: Proceedings of 2018 The 2nd International Conference on Machine Learning and Soft Computing ICMLSC 2018, ACM ISBN: 978–1–4503–6336–5, pp. 57–61 (2018).

---

[1] https://cran.r-project.org/web/packages/recommenderlab/index.html.

6. Abdel-Basset, M., Gamal, A., Son, L.H., Smarandache, F.: A bipolar neutrosophic multi criteria decision making framework for professional selection. Appl. Sci. **10**(4), 1202 (2020)
7. Selvachandran, G., et al.: A new design of mamdani complex fuzzy inference system for multi-attribute decision making problems. IEEE Trans. Fuzzy Syst. (2019). https://doi.org/10.1109/TFUZZ.2019.2961350
8. Ngan, R.T., Son, L.H., Ali, M., Tamir, D.E., Rishe, N.D., Kandel, A.: Representing complex intuitionistic fuzzy set by quaternion numbers and applications to decision making. Appl. Soft Comput. **87**, 105961 (2020)
9. Manel, A.M., Dongalel, T.D., Bapat, M.S.: Application of fuzzy measure and fuzzy integral in students failure decision making. IOSR J. Math. (IOSR-JM), e-ISSN: 2278–5728, p-ISSN: 2319–765X. **10**(6), Ver. III, pp. 47–53 (2014).
10. Christophe, L., Michel, G.. The Choquet integral for the aggregation of interval scales in multicriteria decision making (2008). arXiv:0804.11762v1
11. Milica, S.: Fuzzy measure identification for multicriteria decision making, eRAF J. Comput. **(5)** (2013).
12. Giang, N.L., et al.: Novel incremental algorithms for attribute reduction from dynamic decision tables using hybrid filter-wrapper with fuzzy partition distance. IEEE Trans. Fuzzy Syst. (2019). https://doi.org/10.1109/TFUZZ.2019.2948586
13. Son, L.H.: A new representation of intuitionistic fuzzy systems and their applications in critical decision making. IEEE Intell. Syst. **35**, 6–17 (2019) https://doi.org/10.1109/MIS.2019.2938441
14. Mingang, C., Pan, L.: Performance evaluation of recommender systems. Int. J. Performability Eng. **13**(8), 1246–1256 (2017). https://doi.org/10.23940/ijpe.17.08.p7.12461256
15. László, K.: k Nearest Neighbors algorithm (kNN). T-61.6020 Special Course in Computer and Information Science, Helsinki Univ. Technol. (2008).
16. Anath, R.K., Maznah, M K., Engku, M.N.E., Abu, B.: A Short survey on the usage of choquet integral and its associated, fuzzy measure in multiple attribute analysis. In: International Conference On Computer Science And Computational Intelligence - ICCSCI Procedia Comput.Sci. **59**, pp. 427–434 (2015).
17. Dat, L.Q., et al.: Linguistic approaches to interval complex neutrosophic sets in decision making. IEEE access **7**, 38902–38917 (2019)
18. Thong, N.T., Dat, L.Q., Son, L.H., Hoa, N.D., Ali, M., Smarandache, F.: Dynamic interval valued neutrosophic set: modeling decision making in dynamic environments. Comput. Ind. **108**, 45–52 (2019)
19. Tri, H.M., Vu, T.T., Hung, H.H., Hiep, X.H.: Solution for ordered weighted averaging operator for making in the interaction multicriteria decision in user-based collaborative filtering recommender system. Int. J. Mach. Learn. Comput. (IJMLC) **8**(4) (2018).
20. Tri, H.M., Vu, T.T., Hung, H.H., Hiep, X H.: Item-Based Collaborative Filtering In The Multi-Criteria Recommender System With Ordered Weighted Averaging Operator On Sparse Datasets. JP J. Heat Mass Transf. Special Volume, Issue II, Advances in Mechanical System and ICT-convergence, 183–194 (2018) https://doi.org/10.17654/HMSI218183
21. Ferdaous, H., Bouchra, F., Brahim, O., Ismail, K.: Multi-criteria recommender systems: a survey and a method to learn new user's profile. Int. J. Mob. Comput. Multimedia Commun. **8**(4), 20–48 (2017)
22. Abderrahmane, K., Omar, N., Mohammad, Y.H.A.: A multicriteria collaborative filtering recommender system using learning-to-rank and rank aggregation. Arab. J. Sci. Eng. 1–11 (2019).
23. Umaia, M.A., Shahrul, A.M.N.: Multi-criteria review-based, recommender system—the state of the art. published into the digital object identifier, IEEE Access, **7**, 169446-169468 (2019). https://doi.org/10.1109/ACCESS.2019.2954861

Application of Machine Learning Techniques to Classify Intention to Pay for Forest Ecosystem Services

Pham Thu Thuy[1,2]([⊠]), Nguyen Thanh Tung[3], and Luu Quoc Dat[4]

[1] VNU School of Interdisciplinary Studies, Vietnam National University, 144 Xuan Thuy Road, Hanoi 100000, Vietnam
phamthuthuy@vnu.edu.vn
[2] Science and Technology Department, Vietnam National University, 144 Xuan Thuy Road, Hanoi 100000, Vietnam
[3] International School, Vietnam National University, 144 Xuan Thuy Road, Hanoi 100000, Vietnam
tung_nt@vnu.edu.vn
[4] VNU University of Economics and Business, Vietnam National University, 144 Xuan Thuy Road, Hanoi 100000, Vietnam

Abstract. Capturing the ability to take part in the payment of forest ecosystem services by beneficiaries is the result that policy-making agencies are always concerned. This research selects several machine learning techniques, including single classifiers (Multilayer Perceptron, Naive Bayes, SMO) and ensemble classifiers (LogitBoost, Random Forest, Bagging) to evaluate and classify willingness-to-pay intention for mangrove ecosystem services of people in PhuLong commune, Vietnam. Research data is inherited from a previous contingent valuation survey, with a sample size of 235. The results show that the machine learning algorithms are workt with small sample-size data sets with feasibility prediction results in behavioral intent classification. The LogitBoost model achieves the best classification performance compared to the remaining models. Besides, socio-psychological factors are ranked as important factors in classifying behavioral intentions related to payment for forest ecosystem services.

Keywords: Intention to pay · forest ecosystem services · classification · machine learning

1 Introduction

Forest ecosystems provide public goods characterized by non-excluded and non-competitive nature (Tóth et al., 2010). The value of ecosystem services is often determined through hypothetical markets (established in the Contingent valuation method) rather than traditional markets (Asmare et al., 2022; Obeng Aguilar, 2021; Getachew, 2018). There is ample evidence to show that climate change and economic activities

P. Cong Vinh and N. Thanh Tung (Eds.): ICCASA 2023, LNICST 579, pp. 54–69, 2024.
https://doi.org/10.1007/978-3-031-58878-5_4

to maximize forest-related benefits such as logging, land use conversion, and forest product exploitation is a major causes of the decline in the quality and quantity of forest ecosystem services worldwide (Roesch Rabotyagov, 2016; Taylor et al., 2015; Van, 2012; Raudsepp-Hearne et al., 2010).

In climate change, strengthening forest ecosystem services and the potential exploitation of forests has become a global concern in recent decades (IPCC, 2022; Pagiola et al., 2004; Edward, 2004). According to Obeng Aguilar (2018), when implementing ecosystem service payment programs, it is necessary to classify and evaluate the public's readiness (willingness to pay) for these programs to ensure the financial viability of payment programs.

Many statistical analysis methods have been used to identify the critical factors that affect respondents' decisions about payment intent related to the environment in a multivariate regression model (State et al., 2016; Mojiol et al., 2022), Logit regression model (Pham et al., 2018; Pouta Rekola, 2001), Tobit model (Asmare et al., 2022; Tran et al., 2002; Pouta Rekola, 2001), Structured models (Li et al., 2015; Lin Syrgabayeva, 2016; Liu et al., 2020). According to Sokratous et al. (2023), probabilistic models come with a wide range of problems stemming from reality, (i) many types of behavior can arise from stochastic processes, (ii) many theoretical models have never been explored because they do not have convenient probability density functions for different combinations of parameters, (iii) it is challenging to combine all the variables from many questions to determine conditional probability.

Machine learning models began to be approached as an alternative in several studies related to pay preferences (Sokratous et al., 2023). Phan et al. (2021) have adopted a logistic regression model and Bayes network to identify determinants that affect willingness to pay (willingness to pay) for reservoir construction, increasing water prices. Subhan et al. (2023) also used machine learning and quantum regression approaches to research willingness to pay to enhance road safety. Ayansola et al. (2022) used K-Nearest Neighbors, Random Forest, Support Vector Machine Decision Tree, and Boosting to model consumers' willingness to pay for electricity.

Machine learning methods can use data from a variety of sources or integrate survey questions into models without transforming or combining survey data (Shen et al., 2021; Thuy et al., 2021). Besides, when applied to a new situation that has never appeared, the trained model will make predictions using the trained generalization samples (Crook et al., 2007; Sokratous et al., 2023). Many studies have shown that machine learning models can provide information with higher accuracy and reliability than traditional regression models (Bashar et al., 2023; Zeng et al., 2019; Selz, 2020).

Machine learning algorithms often prioritize large-sample-size data sets due to the growing capacity of computing systems; the more significant the data size, the greater the statistical capacity. In addition, recent data collection methods are becoming cheaper and more accessible (Vabalas et al., 2019; Raudys Jain, 1991). On the other hand, studies based on traditional methods, such as contingent valuation (CVM), often require face-to-face interviews. The US National Oceanic and Atmospheric Administration (NOAA) recommends that face-to-face interviews minimize bias when implementing CVM. However this way of implementation often requires significant expense, so large data samples are often not preferred. Therefore, developing machine learning methods to solve problems with small datasets needs further exploration.

Although machine learning methods are increasingly applied to solve environmental management issues, the literature review shows that it needs to be of more use in classifying human preferences for ecosystem services. The objective of this research are, (i) to apply machine learning models to classify people's willingness to pay for mangrove ecosystem services, (ii) discover whether machine learning models give positive results on a small-sample-sizes dataset that collected earlier through CVM, (iii) evaluates the performance results of models with data sets with and without important feature options.

The rest of this study is organized as follows, Sect. 2 proposes the research design. Section 3 presents the research methodology. Section 4 overviews the research data. Section 5 presents about research results. Finally Discussion and Conclusion are drawn in Sect. 6 (Fig. 1).

2   Research Design

2.1   Research Design Diagram

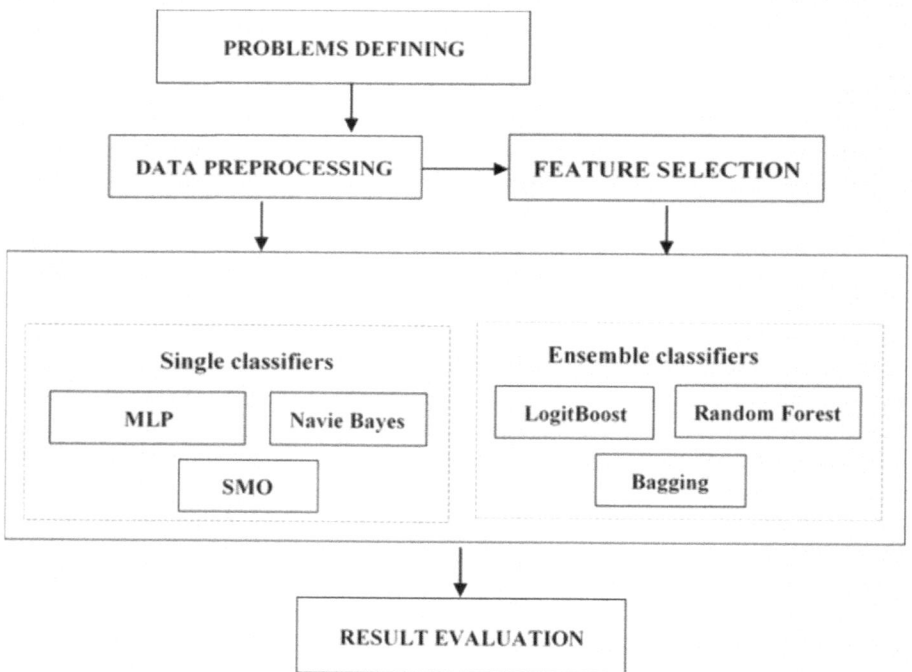

Fig. 1. Research design model (author)

2.2   Implementation Steps

Step 1, Problems defining.

The research problem can be stated as follows:

For dataset X, X = {x_1, x_2, x_3,..., x_n}, n is the number of sample, xn is the vector including the demographic, socio-economic, and psychological characteristics of the nth object. Set of classification labels W, W = { w_1, w_2, w_3, ... w_n}; $w_n \in$ {1; 0}. The problem is giving correct prediction of a label for each sample of dataset.

Step 2, Data Preprocessing.

Data preprocessing involves transforming raw data into a format the machine can comprehend. Some specific fields of information in the dataset might lack meaning or fail to meet the conditions for execution, necessitating removal or normalization.

After preprocessing, the dataset will be cleaned, standardized, and encoded into categories and sets, as the machine learning model requires.

Step 3.1, Features Selection.

Machine learning models often come with diverse datasets in terms of features. Feature selection is necessary to ensure accurate predictive results. This work helps reduce the number of parameters or training time and overfitting issues. It also aims to achieve a balanced classification and establish the best model with high prediction capability and minimal error.

Step 3.2. Intention to pay classification.

In this study, single classifiers and ensemble classifiers were chosen for experimentation. The selected single classifiers include Multilayer Perceptron, Naive Bayes, and SMO (SVO). The selected ensemble classifiers include LogitBoost, Random Forest, and Bagging. The aim was to evaluate and compare the performance of these classifiers in the context of the research.

Step 4, Evaluate and compare attribute classification results and model performance.

2.3   Evaluation Metrics

The evaluation metrics Accuracy, Precision, Recall, and F1-score were chosen to classify and evaluate the performance of the machine learning methods, Accuracy, Precision, Recall, and F1-score. These metrics are widely used and commonly employed (Goutte Gaussier, 2005). However, the relative importance of each metric depends on the specific problem and the associated costs related to classification outcomes for each task. These criteria are measured on a scale from 0 (terrible classification) to 1 (perfect classification) using the following equations,

$$\text{Accuracy} = \frac{\text{Number of true predictions}}{\text{Total number of predictions}}$$

$$\text{Precision} = \frac{\text{TP}}{\text{TP} + \text{FP}}$$

$$\text{Recall} = \frac{\text{TP}}{\text{TP} + \text{FN}}$$

$$F1 - score = \frac{2 \text{ x (Precision x Recall)}}{Precision + Recall}$$

In there, TP, True Positive; TN, True Negative; FP, False Positive; FN, False Negative.

2.4 Experimental Setups

The k-fold cross-validation technique is employed to assess the predictive performance. This approach divides the dataset into k subsets of equal size. One subset is used as validation data, while the remaining (k-1) subsets are used as training data to parameterize the models. The training and testing process of the models is repeated until all k subsets have been used. For this study, the chosen value of k for the models' execution is 10.

2.5 Tools for Experiments

In this study, the Weka software version 3.8.6 has been chosen as the tool to conduct machine learning algorithms. Weka also provides tools to preprocess data and select features in this research.

3 Research Methods

3.1 Single Classifiers Techniques

3.1.1 Sequential Minimal Optimization (SMO)

Sequential Minimal Optimization (or SMO) is an algorithm used for training Support Vector Machine (SVM), a type of machine learning model commonly used for classification and regression tasks (Liang et al., 2017; Naik Desai, 2017). SMO is designed to efficiently solve the optimization problem of training an SVM, which involves finding a solution that maximizes the margin between different classes while minimizing classification errors (Platt, 1998). SMO breaks this problem into a series of smallest possible sub-problems, which are then solved analytically. Because of the linear equality constraint involving the Lagrange multipliers αi, the smallest possible problem involves two such multipliers. Then, for any two multipliers α_1 and α_2, the constraints are reduced to,

$$0 \leq \alpha_1, \alpha_2 \leq C$$

$$y_1 a_1 + y_2 a_2 = k$$

Moreover, this reduced problem can be solved analytically, one must find a minimum of a one-dimensional quadratic function. k is the negative of the sum over the rest of the terms in the equality constraint, which is fixed in each iteration. The algorithm proceeds as follows,

- Find a Lagrange multiplier $\alpha 1$ that violates the Karush–Kuhn–Tucker (KKT) conditions for the optimization problem.
- Pick a second multiplier $\alpha 1$ and optimize the pair (α_1, α_2).

- Repeat steps 1 and 2 until convergence.

When all the Lagrange multipliers satisfy the KKT conditions (within a user-defined tolerance), the problem has been solved. Although this algorithm is guaranteed to converge, heuristics are used to choose the pair of multipliers to accelerate the convergence rate, and this is critical for large data sets since a number p represents the possible choices for α_i and α_j. This number p is calculated by this formula,

$$p = \frac{n(n-1)}{2}$$

Although SMO was designed for SVM training, it is worth mentioning that there are other optimization techniques and libraries available that can train SVMs effectively as well. Nonetheless, SMO remains a foundational algorithm in the SVM literature and has contributed to the widespread use of SVMs in various machine-learning applications (Noronha et al., 2019).

3.1.2 Multilayer Perceptron (MLP)

A Multilayer Perceptron is an artificial neural network consisting of multiple layers of interconnected nodes, or "neurons". It is a fundamental architecture used in machine learning and deep learning for various tasks such as classification, regression, and pattern recognition (Botalb et al., 2018). Neurons within an MLP are organized feedforward, meaning the data flows from the input layer through the hidden layers to the output layer without forming any cycles. Each connection between neurons has an associated weight that the network adjusts during training to optimize its performance on the given task.

Training an MLP involves feeding it with labeled data (input-output pairs) and using a process called backpropagation to adjust the weights based on the difference between predicted and target outputs. This iterative optimization process aims to minimize a chosen loss function, which quantifies the difference between predictions and ground truth. The two historically common activation functions are both sigmoid functions which are described by two formulas,

$$y(v_i) = \tanh(v_i)$$

$$y(v_i) = \left(1 + e^{-v_i}\right)^{-1}$$

MLPs are considered shallow neural networks compared to more complex architectures like convolutional neural networks (CNNs) and recurrent neural networks (RNNs). However, they can still model various functions and have been used successfully in various applications.

3.1.3 Naive Bayes

Naive Bayes is a probabilistic machine learning algorithm for classification and some-times regression tasks. It is based on Bayes' theorem and the assumption of conditional independence among features given the class label (Chaudhary et al., 2016). Despite the simplifying assumption of independence, Naive Bayes can be surprisingly effective for many real-world problems. At the core of Naive Bayes is Bayes' theorem, which calcu-lates the probability of a hypothesis (class label) given the observed evidence (features). It is expressed as the below mathematical formula,

$$p\,(C_k \mid x) = \frac{p\,(C_k) * p\,(x \mid C_k)}{p\,(x)}$$

The fraction's numerator plays a vital role in the equation because of the independence between C and the value of feature x_i. It equals to model $p\,(C_k, x_1, \ldots, x_n)$. Thus, this equation was rewritten using the chain rule for repeated applications of the definition of conditional probability.

$$
\begin{aligned}
p(C_k, X_1, \ldots, X_n) &= p(X_1, X_2, \ldots, X_n, C_k) \\
&= p(X_1 \mid X_2, \ldots, X_n, C_k).p(X_2, \ldots, X_n, C_k) \\
&= p(X_1 \mid X_2, \ldots, X_n, C_k).p(X_2 \mid X_3, \ldots, X_n, C_k).p(X_3, \ldots, X_n, C_k) \\
&= \ldots \\
&= p(X_1 \mid X_2, \ldots X_n, C_k).\ p(X_2 \mid X_3, \ldots, X_n, C_k).\ldots.p(X_n - 1 \mid X_n, C_k) \\
&\quad .\ p(X_n \mid C_k).\ p(Ck)
\end{aligned}
$$

Naive Bayes has limitations, mainly when the independence assumption does not hold actual or features are highly correlated (Saritas Yasar, 2019). However, its simplicity, efficiency, and ability to perform well on specific tasks make it a valuable tool in many machine-learning applications.

3.2   Ensemble Classifiers Techniques

3.2.1   Logit Boost

Logit Boost is an ensemble machine-learning algorithm that belongs to the boosting fam-ily of techniques. Boosting is a method that combines multiple weak learners (typically simple models) into strong learners, iteratively focusing on the instances that the current ensemble is struggling to classify correctly (Tehrany et al., 2019). (Tehrany et al., 2019). Logit Boost is designed for binary classification tasks. Logit Boost can be interpreted as a convex optimization process. To elaborate, when we aim to obtain an additive model in the formula (Dettling Bühlmann, 2003),

$$f = \sum_t \alpha_t h_t$$

Thus, the LogitBoost algorithm reduces the logistic loss to the lowest value,

$$\sum_i \log\left(1 + e^{-y_i f(x_i)}\right)$$

Work Process of LogitBoost,

- Initialization, Initialize the training data weights. Initially, all data points are given equal weights.
- Iterative Process, LogitBoost performs a series of iterations
- Final Prediction, The final prediction is made by combining the predictions of all weak learners in the ensemble, weighted by their corresponding boosting coefficients.

LogitBoost is a robust algorithm that has demonstrated exemplary performance in various applications. However, tuning hyperparameters is essential to avoid overfitting and achieve optimal results like other boosting methods.

3.2.2 Random Forest

Random Forest is a powerful ensemble learning algorithm used for classification and regression tasks in machine learning (Chaudhary et al., 2016). It is particularly effective for handling complex datasets and producing accurate predictions. The algorithm belongs to the family of ensemble methods, which combine the predictions of multiple individual models to create a more robust overall model (Liu et al., 2012).

The initial step in assessing variable importance within dataset A is expressed by the formula below, First, a random forest is fitted to the dataset. During this fitting process, the out-of-bag error is calculated for each data point and averaged across the entire forest. If bagging is not employed during training, this can be replaced with errors from an independent test set.

$$D_n = \{X_i, Y_i\}_{i-1}^n$$

To gauge the importance of feature rank j post-training, the values corresponding to the jth feature rank are permuted across the training data. Subsequently, the out-of-bag error is recalculated on this perturbed dataset. The critical score for feature rank j is computed by taking the average of the differences in out-of-bag error before and after the permutation across all trees. This score is then normalized using the standard deviation of these computed differences.

Random Forests are widely used in practice and have found applications in various fields, including finance, medicine, marketing, and more. They are relatively easy to use and require fewer tunable hyperparameters than other complex algorithms. The only primary parameter that typically needs to be adjusted is the number of trees in the ensemble.

3.2.3 Bagging

Bagging, short for "Bootstrap Aggregating", is an ensemble learning technique used to improve the accuracy and robustness of machine learning models, especially decision trees (Breiman, 1996). It involves training multiple instances of the same model on different subsets of the training data and then combining their predictions to make a final prediction (Yaman Subasi, 2019). The goal of bagging is to reduce overfitting and increase the model's generalization performance. The training process for random forests employs the fundamental method of bootstrap aggregating, often called bagging, in the context of tree learners. In the presence of a training set $X = x_1,..., x_n$ with corresponding responses $Y = y_1,..., y_n$, the bagging procedure involves iteratively selecting a random sample from the training set, employing replacement, and then constructing trees based on these samples. This process is repeated B times, and for each iteration $b = 1,...,$ B, the process begins by initially selecting a random sample, with replacement, of n training examples from the sets X and Y. This sampled subset is denoted as X_b and Y_b. Subsequently, a classification or regression tree f_b is trained using the data from X_b and Y_b. Following the training phase, predictions for previously unseen samples x' are generated by averaging the predictions derived from all individual regression trees on x',

$$\hat{f} = \frac{1}{B} \sum_{b=1}^{B} f_B(x')$$

Bagging is not limited to decision trees but can also be applied to other base models (Yaman Subasi, 2019). As previously mentioned, Random Forest is a specific ensemble method that uses bagging with decision trees as its base models. Bagging can also be a foundation for more advanced ensemble techniques like boosting. Overall, bagging is a powerful tool to enhance the performance and reliability of machine learning models.

4   Research Data

The research data is taken from the survey results of households living in Phu Long commune, Hai Phong City, Vietnam, in 2022. The survey questionnaire is designed according to the Contingent valuation method with this dataset. Respondents in the survey are heads of households. In the interview scenario, respondents were first provided information about the status of mangroves. After that, they were asked to answer questions to explore the intention to pay (willingness to pay) for benefiting from mangrove ecosystem services through the forest payment project.

The database has a sample size of 235. Demographic, socio-psychological factors, and socio-economic characteristics classify intention to pay. The data structure is summarized in Table 1. The data fields are normalized and labeled to ensure the execution conditions of machine learning algorithms. One hundred eighty (180) samples are labeled willing to pay with the value "yes", and fifty-five (55) samples are labeled unwilling to pay with the value "no" which are encoded as 1 and 0, respectively.

Table 1. Data structure

Variables symbol	Variables name	Measurement
Socio-psychological factors		The mean value of observed variables
Att	Attitude	Real
Sjn	Subjective norms	Real
Pbc	Perceived behavioral control	Real
Knl	Knowledge	Real
Wtp (*predictor variable*)	Willingness to pay for forest environment service	{1; 0}
Socio-economic characteristics		
Age	Age of household head	Natural number
lnIncome	Average household income/month (million VietNam Dong)	Real
Edu	Education level of household head	1, Primary school 2, Secondary school 3, High school 4, College 5, University 6, Postgraduate
Gender	Gender of household head	1, Male 2, Female
Job	The main occupation of the household	1, Raising, planting, and exploiting seafood; 2, Other
Hhsize	Number of people in the household	Natural number

5 Result

Feature selection is crucial in determining relevant features from a dataset for classification and prediction purposes. The classification performance is expected to improve when the dataset contains predictive variables with high-value contributions. This research uses two technicals to feature selection, Information Gain and Correlation Attribute Ranking Filters. From the original dataset, five attributes that significantly influence the dependent variable are ranked in descending order of importance, as shown in Table 2. The results indicate that most of the socio-psychological variables play a dominant role in machine learning models for classifying willingness-to-pay intention. Only the "income" variable holds relative importance in the model among the demographic and socio-economic variables.

Table 2. Ranking values of Information Gain and Correlation Attributes Evaluator

Attributes	Information Gain ranking values	Correlation Attributes ranking values
Att	0.250	0.579
Knl	0.209	0.484
Sjn	0.098	0.366
Pbc	0.072	0.327
Income	0.053	0.274
Job	0.004	0.171
Gender	0.002	0.085
Hhsize	0	0.081
Edu	0	0.052
Age	0	0.048

Table 3 shows the result of using the full 11 parameters (including age, edu, gender, hhsize, job, pbc, att, knl, sjn, income, wtp) for machine learning models. In this case, all six selection algorithms obtained relatively good accuracy ($\geq 80\%$). In the Ensemble classifiers group, the LogitBoost model gave a significantly better accuracy than Bagging and Random Forest. The values achieved were 87.234%, 83.404%, and 87.234%, respectively.

Table 3. Experimental results of classification of willingness to pay using machine learning algorithms with full input parameters

Machine learning methods	Accuracy	Precision	Recall	F1-Score
Single classifiers				
Naive Bayes	84.255%	0.838	0.843	0.840
SMO	84.680%	0.840	0.840	0.830
Multilayer Perceptron	80.000%	0.792	0.800	0.795
Ensemble classifier				
LogitBoost	87.234%	0.867	0.872	0.865
Random Forest	84.255%	0.835	0.843	0.836
Bagging	83.404%	0.824	0.834	0.824

Table 4 shows the results of conducting machine learning algorithms with six (06) essential parameters selected by two technicals for feature selection before (including pbc, att, knl, sjn, income, and WTP). After reducing the input parameters, the algorithms in the single classifiers group tended to increase performance, while the ensemble classifiers group tended to differ from the three (3) algorithms.

After reducing the input parameters, the algorithms in the single classifiers group increased accuracy, while the ensemble classifiers group tended to differ from the three algorithms. Multilayer Perceptron has the most impressive change (from 80% to 85.1%). On the contrary, Random Forest is the only algorithm out of the six models with an accuracy reduction (84.255% to 82,128%). The LogitBoost model is almost unaffected regarding results when reducing the number of model input parameters.

Table 4. Experimental results of classification of willingness to pay using machine learning algorithms and important attribute selection

Machine learning methods	Accuracy	Precision	Recall	F1-Score
Single classifiers				
Naive Bayes	84.680%	0.845	0.847	0.846
SMO	85.957%	0.856	0.860	0.847
Multilayer Perceptron	85.106%	0.843	0.851	0.844
Ensemble classifier				
LogitBoost	87.234%	0.867	0.872	0.865
Random Forest	82.128%	0.819	0.821	0.820
Bagging	83.405%	0.824	0.834	0.824

In both cases, LogitBoost is a model for better results than other methods for all evaluation indicators, details shown in Table 5. The correct accuracy reached 87.234%, and the F1-Score values for classes "1" and "0" were 0.920 and 0.688 respectively. The Precision and Recall scores for the case "willing to pay" is relatively high, respectively 0.865 and 0.956. However, for the case of "not willing to pay", the Precision and Recall scores have relative differences. The values are 0.805 and 0.600, respectively.

Table 5. LogitBoost model experimental results (best)

Machine learning methods	Accuracy		Class	Precision	Recall	F1-Score
	Correctly	Incorectly				
LogitBoost	87.234%	12.766%	1	0.887	0.956	0.920
			0	0.805	0.600	0.688
			ave	0.867	0.872	0.865

6 Discussion and Conclusion

The machine learning algorithms tested in this study provided workable prediction results for classifying willingness-to-pay intentions for mangrove ecosystem services. This outcome enhances comparability and confidence in determining strategies for payment programs related to the mangrove forest. Experimental results show that LogitBoost achieved the best performance among the models regarding accuracy and precision when tested with the research dataset.

Changing the number of attributes has a different effect on performing the tested machine-learning models. Notably, most models improved performance, with Multilayer Perceptron exhibiting the most impressive classification enhancement. Conversely, Random Forest was the only model to experience a decrease in performance when keeping only the more critical variables. LogitBoost remained unaffected by changes in the number of input attributes.

Psychological factors were ranked higher in importance than demographic and socioeconomic factors in classifying willingness-to-pay intentions. Four out of five independent variables ranked as important and significant for the classification model belonged to the psychological group. This underscores that willingness to pay is a behavioral intention that needs to be examined based on socio-psychological contexts. However, in this small-sample dataset, the limited number of instances labeled as "not willing to pay" (55) likely contributed to the relative difference in Precision and Recall scores for this class. Therefore, combined and enhanced techniques are recommended to further explore to improve prediction performance of the models.

7 Declaration of Competing Interest.

The authors declare that they have no known competing financial interests or personal relationships that could have appeared to influence the work reported in this paper.

References

1. Asmare, E., Bekele, K., Fentaw, S.: Households willingness to pay for the rehabilitation of wetlands, evidence from Gudera Wetland, Northwest Ethiopia. Heliyon **8** (2022). https://doi.org/10.1016/j.heliyon.2022.e08813
2. Ayansola, O., Ogundunmade, T., Adedamola, A.: Modelling willingness to pay of electricity supply using machine learning approach. Modern Econ. Manag. 0–6 (2022). https://doi.org/10.53964/mem.2022009
3. Bashar, M.A., Nayak, R., Astin-Walmsley, K., Heath, K.: Machine learning for predicting propensity-to-pay energy bills. Intell. Syst. Appl. **17**, 200176 (2023). https://doi.org/10.1016/j.iswa.2023.200176
4. Botalb, A., Moinuddin, M., Al-Saggaf, A.S.: Contrasting convolutional neural network (CNN) with multi-layer perceptron (MLP) for big data analysis. In: 2018 International Conference on Intelligent and Advanced System (ICIAS), Kuala Lumpur, Malaysia, pp. 1–5 (2018) https://doi.org/10.1109/ICIAS.2018.8540626
5. Breiman, L.: Bagging predictors. Mach. Learn. **24**(2), 123–140 (1996). https://doi.org/10.1007/BF00058655

6. Chaudhary, A., Kolhe, S., Kamal, R.: An improved random forest classifier for multi-class classification. Inform. Process. Agri. **3**(4), 215–222 (2016). https://doi.org/10.1016/j.inpa. 2016.08.002

7. Crook, J., Edelman, D., Thomas, L.: Recent developments in Consumer Credit Risk assessment. Eur. J. Oper. Res. **183**, 1447–1465 (2007). https://doi.org/10.1016/j.ejor.2006. 09.100

8. Dettling, M., Bühlmann, P.: Boosting for tumor classification with gene expression data. Bioinformatics **19**(9), 1061–1069 (2003). https://doi.org/10.1093/bioinformatics/btf867

9. Edward, B.B., Suthawan, S.: Shrimp Farming and Mangrove Loss in Thailand. In: Edward B.B (ed.), Elgar Publisher, Chulalongkorn University, Thailand (2004)

10. Getachew, T.: Estimating willingness to pay for forest ecosystem conservation the case of Wof-Washa forest, North Shewa Zone, Amhara National regional state, Ethiopia. J. Res. Develop. Manag. **46**, 46–61 (2018)

11. Goutte, C., Gaussier, E.: A probabilistic interpretation of precision, recall and F-score, with implication for evaluation. In: Losada, D.E., Fernández-Luna, J.M. (eds.) Advances in Information Retrieval. ECIR 2005. Lecture Notes in Computer Science, vol. 3408, pp. 345–359. Springer, Heidelberg (2005). https://doi.org/10.1007/978-3-540-31865-1_25

12. IPCC: Climate change 2022, impacts, adaptation, and vulnerability. In: Contribution of Working Group II to the Sixth Assessment Report of the Intergovernmental Panel on Climate Change. Cambridge University Press, Cambridge, UK and New York. https://doi.org/10. 1017/9781009325844

13. Li, Y., Deng, H., Dong, R.: Prioritizing protection measures through ecosystem services valuation for the Napahai Wetland, Shangri-La County, Yunnan Province, China. Int J Sust Dev World **22**(2), 142–150 (2015). https://doi.org/10.1080/13504509.2014.926298

14. Liang, X., Zhu, L., Huang, D.S.: Multi-task ranking SVM for image cosegmentation. Neurocomputing, **247**, 126–136 (2017). https://doi.org/10.1016/j.neucom.2017.03.060

15. Lin, C.Y., Syrgabayeva, D.: Mechanism of environmental concern on intention to pay more for renewable energy, application to a developing country. Asia Pac. Manag. Rev. **21**(3), 125–134 (2016). https://doi.org/10.1016/j.apmrv.2016.01.001

16. Liu, P., Teng, M., Han, C.: How does environmental knowledge translate into pro-environmental behaviors? The mediating role of environmental attitudes and behavioral intentions. Sci. Total. Environ. **728**, 138126 (2020). https://doi.org/10.1016/j.scitotenv.2020. 138126

17. Liu, Y., Wang, Y., Zhang, J.: New machine learning algorithm: Random Forest. In: Liu, B., Ma, M., Chang, J. (eds.) Information Computing and Applications, ICICA 2012. LNCS, vol. 7473, pp. 246–252. Springer, Heidelberg (2012). https://doi.org/10.1007/978-3-642-34062-8_32

18. Mojiol, A.R., Hong, K.Y., Saleh, E.: Willingness to pay for mangroves conservation by the local communities in Salut Mengkabong Lagoon Tuaran Sabah. Hutan Tropika, **14**(1), 1–9 (2022). https://doi.org/10.36873/jht.v14i1.310

19. Naik, V., Desai, A.: Online handwritten Gujarati character recognition using SVM, MLP, and K-NN. In: 8th International Conference on Computing, Communication and Networking Technologies, 1–6 (2017). https://doi.org/10.1109/ICCCNT.2017.8203926

20. Noronha, D.H., Torquato, M.F., Fernandes, M.A.C.: A parallel implementation of sequential minimal optimization on FPGA. Microprocess. Microsyst. **69**, 138–151 (2019). https://doi. org/10.1016/j.micpro.2019.06.007

21. Obeng, E.A., Aguilar, F.X.: Value orientation and payment for ecosystem services, Perceived detrimental consequences lead to willingness-to-pay for ecosystem services. J. Environ. Manage. **206**, 458–471 (2018). https://doi.org/10.1016/j.jenvman.2017.10.059

22. Obeng, E.A., Aguilar, F.X.: Willingness-to-pay for restoration of water quality services across geo-political boundaries. Curr. Res. Environ. Sustain. **3**, 100037 (2021). https://doi.org/10.1016/j.crsust.2021.100037

23. Pagiola, S., Ritter, K., Bishop, J.: How Much is an Ecosystem Worth? Assessing the Economic Value of Conservation (2004)

24. Pham, T.D., Kaida, N., Yoshino, K., Nguyen, X.H., Nguyen, H.T., Bui, D.T.: Willingness to pay for mangrove restoration in the context of climate change in the Cat Ba biosphere reserve Vietnam. Ocean Coast. Manag. **163**, 269–277 (2018). https://doi.org/10.1016/j.ocecoaman.2018.07.005

25. Phan, T.D., Bertone, E., Pham, T.D., Pham, T.V.: Perceptions and willingness to pay for water management on a highly developed tourism island under climate change, a Bayesian network approach. Environ. Challenges **5**, 100333 (2021). https://doi.org/10.1016/j.envc.2021.100333

26. Platt, J.: Fast training of support vector machines using sequential minimal optimization. Adv. kernel Methods - Support Vector Learn. (1998). https://www.microsoft.com/en-us/research/publication/fast-training-of-support-vector-machines-using-sequential-minimal-optimization/

27. Pouta, E., Rekola, M.: The theory of planned behavior in predicting willingness to pay for abatement of forest regeneration. Soc. Nat. Resour. **14**(2), 93–106 (2001). https://doi.org/10.1080/089419201300000517

28. Raudsepp-Hearne, C., Peterson, G., Bennett, E.: Ecosystem service bundles for analyzing tradeoffs in diverse landscapes. Proc. Natl. Acad. Sci. U.S.A. **107**, 5242–5247 (2010). https://doi.org/10.1073/pnas.0907284107

29. Raudys, S.J., Jain, A.K.: Small sample size effects in statistical pattern recognition, recommendations for practitioners. IEEE Trans. Pattern Anal. Mach. Intell. **13**(3), 252–264 (1991). https://doi.org/10.1109/34.75512

30. Roesch-McNally, G.E., Rabotyagov, S.S.: Paying for forest ecosystem services, voluntary versus mandatory payments. Environ. Manage. **57**(3), 585–600 (2016). https://doi.org/10.1007/s00267-015-0641-7

31. Saritas, M.M., Yasar, A.: Performance analysis of ANN and Naive Bayes classification algorithm for data classification. Int. J. Intell. Syst. Appl. Eng. **7**(2), 88–91 (2019). https://doi.org/10.18201//ijisae.2019252786

32. Selz, D.: From electronic markets to data driven insights. Electron. Mark. **30**(1), 57–59 (2020). https://doi.org/10.1007/s12525-019-00393-4

33. Shen, Y., Hamm, J.A., Gao, F., Ryser, E.T., Zhang, W.: Assessing consumer buy and pay preferences for labeled food products with statistical and machine learning methods. J. Food Prot. **84**(9), 1560–1566 (2021). https://doi.org/10.4315/JFP-20-486

34. Sokratous, K., Fitch, A.K., Kvam, P.D.: How to ask twenty questions and win, Machine learning tools for assessing preferences from small samples of willingness-to-pay prices. J. Choice Model. **48**, 100418 (2023). https://doi.org/10.1016/j.jocm.2023.100418

35. State, D., Ogeh, K.T., Jimoh, S.O., Ajewole, O.I.: Willingness to pay for environmental service functions of mangrove forest. J. Res. Dev. Manag. **16**, 1–7 (2016)

36. Subhan, F., Ali, Y., Zhao, S.: Unraveling preference heterogeneity in willingness-to-pay for enhanced road safety, a hybrid approach of machine learning and quantile regression. Accid. Anal. Prev. **190**, 107176 (2023). https://doi.org/10.1016/j.aap.2023.107176

37. Taylor, P., Bhatta, L.D., Eric, B., Oort, H., Van Stork, N.E., Baral, H.: Ecosystem services and livelihoods in a changing climate, understanding local adaptations in the Upper Koshi, Nepal. Int. J. Biodivers. Sci. Ecosyst. Serv. Manag. 37–41 (2015). https://doi.org/10.1080/21513732.2015.1027793

38. Tehrany, M.S., Jones, S., Shabani, F., Martínez-Álvarez, F., Tien Bui, D.: A novel ensemble modeling approach for the spatial prediction of tropical forest fire susceptibility using logit-boost machine learning classifier and multi-source geospatial data. Theoret. Appl. Climatol. **137**(1–2), 637–653 (2019). https://doi.org/10.1007/s00704-018-2628-9

39. Thuy, P.T., Thanh Tung, N., Hoang, C.D.: Predicting the level of hypertension using machine learning BT. In: Vinh, P.C., Rakib, A. (eds.) Context-Aware Systems and Applications, and Nature of Computation and Communication, pp. 113–122. Springer, Heidelberg (2021). https://doi.org/10.1007/978-3-319-29236-6

40. Tóth, S., Ettl, G., Rabotyagov, S.: ECOSEL, an auction mechanism for forest ecosystem services. MCFNS **2**, 99–116 (2010)

41. Tran, Y.L., Siry, J.P., Bowker, J.M., Poudyal, N.C.: Atlanta households' willingness to increase urban forests to mitigate climate change. Urban For. Urban Green. (2017). https://doi.org/10.1016/j.ufug.2017.02.003

42. Vabalas, A., Gowen, E., Poliakoff, E., Casson, A.J.: Machine learning algorithm validation with a limited sample size. PLoS ONE **14**(11), 1–20 (2019). https://doi.org/10.1371/journal.pone.0224365

43. Yaman, E., Subasi, A.: Comparison of bagging and boosting ensemble machine learning methods for automated EMG signal classification. Biomed. Res. Int. **2019**, 9152506 (2019). https://doi.org/10.1155/2019/9152506

44. Zeng, M., Cao, H., Chen, M., Li, Y.: User behaviour modeling, recommendations, and purchase prediction during shopping festivals. Electron. Mark. **29**(2), 263–274 (2019). https://doi.org/10.1007/s12525-018-0311-8

Anomaly Detection in Univariate Time Series: HOT SAX vesus LSTM-Based Method

Duong Tuan Anh[1,2(✉)] and Tran Long Hoai[2]

[1] Department of Information Technology, HCMC University of Foreign Languages and Information Technology, Ho Chi Minh City, Vietnam
hdt@huflit.edu.vn

[2] Faculty of Computer Science and Engineering, Ho Chi Minh City University of Technology, Ho Chi Minh City, Vietnam
tlhoai.sdh19@hcmut.edu.vn

Abstract. Anomaly detection in time series has been an important and challenging research topic. There have been several methods proposed for detecting anomaly subsequences in a time series. The majority of these methods is classified into the window-based category, which applies a sliding window with a fixed length to extract subsequences before finding out anomaly subsequences. A well-known algorithm in this category is HOT SAX algorithm. Recently, deep neural network models, especially Long Short Term Memory (LSTM) network, are also applied for time series anomaly discovery. LSTM-based methods for time series anomaly detection belong to prediction-based category. So far, there has been no research work to compare the performance of LSTM-based method to that of any traditional window-based method in time series anomaly detection. The research question investigated in this paper is that whether the newly developed LSTM-based method for time series anomaly detection is superior to the traditional algorithms, such as HOT SAX or not. In this study, we give an empirical comparison between LSTM-based method and HOT SAX in time series anomaly detection. Extensive experiments on seven benchmark time series indicate that LSTM-based method is not superior to HOT SAX since it brings out the same detection accuracy as HOT SAX while it incurs much higher computational overhead.

Keywords: Time series · Anomaly detection · Prediction-based approach · Window-based approach · Long Short Term Memory · HOT SAX

1 Introduction

Time series anomaly detection is important in several application areas such as fault detection, disease diagnosis, event detection and data cleaning. For univariate time series there are two commonly used categories of anomaly detection methods: window-based and prediction-based. In a method of window-based category, a sliding window with fixed length w slides through a time series. For each subsequence extracted under the sliding window, the Euclidean distance from it to the closest subsequence in the time series

P. Cong Vinh and N. Thanh Tung (Eds.): ICCASA 2023, LNICST 579, pp. 70–84, 2024.
https://doi.org/10.1007/978-3-031-58878-5_5

is computed and used as the anomaly score of the subsequences. Empirical evaluations have shown that this simple method is effective for many different kinds of time series. In a method of prediction-based category, a prediction model is trained to learn the normal behavior of a time series segment (training part of the time series), and prediction errors are used to discover anomaly patterns on the test part of the time series with the rule that any points very different to their predicted values are considered as anomaly points.

Some typical algorithms of the first category are Brute-Force, a naïve method proposed by Keogh et al. (2005) [1]; HOT SAX, devised by Keogh et al. (2005) [1]; WAT algorithm by Bu et al. (2007) [2] and BitClusterDiscord algorithm by Li et al. (2013) [3].

Some typical algorithms of the second category can be listed as follows. Oliveira and Meira, in 2006, proposed a method which can detect anomaly patterns through neural network forecasting with robust confidence intervals [4]. Pena et al. in 2013, proposed a method for time series anomaly detection which is based on ARIMA model and Holt-Winters model [5]. Yu et al. in 2014, proposed a method for time series anomaly detection which is based on AR (Auto Regression) model and sliding window prediction [6].

Recently, deep neural network-based anomaly detection algorithms have become increasingly popular and have been applied for a variety of practical areas. Most of deep neural network-based anomaly detection algorithms belong to prediction-based category. Since LSTM network outperforms many other models in time series forecasting ([7–10]), LSTM-based approach has been used in deep learning-based anomaly detection algorithms which belongs to prediction-based category [11]. Some typical LSTM-based methods in time series anomaly detection can be listed as follows. Malhotra et al., in 2015, proposed a time series anomaly discovery method, called LSTM-AD, which utilizes stacked LSTM for prediction and uses prediction errors to detect anomalies [12]. Chauhan and Vig in 2015 devised an LSTM-based method to detect anomalies in electrocardiography (ECG) time series [13]. Buda et al. in 2018 proposed a time series anomaly discovery method, called DeepAD which combines stacked LSTM with traditional forecasting methods such as ARIMA and Holt-Winters in a prediction-based anomaly detection algorithm [14]. Zhang et al. in 2020 proposed an LSTM-based algorithm which can forecast electrical load along with detecting anomalies and adjusting them in order to improve forecasting quality at real time [15].

An interesting and important research question is whether the newly developed LSTM-based method for time series anomaly detection is superior to the traditional algorithms, such as HOT SAX algorithm or not. To the best of our knowledge, there is no specific empirical research work to assess the performance of LSTM-based method in time series anomaly discovery in comparison with traditional methods such as HOT SAX.

This work aims to compare LSTM-based method and HOT SAX in time series anomaly detection on seven benchmark datasets in two perspectives detection accuracy and time efficiency. Extensive experiments on seven benchmark time series indicate that LSTM-based method is not superior to HOT SAX since it brings out the same detection accuracy as HOT SAX while it incurs much higher computational overhead.

The rest of the paper is structured as follows. Section 2 presents some basic definitions about time series anomalies, taxonomy of time series anomaly detection and

LSTM neural networks. In Sect. 3, we describe the LSTM-based approach and HOT SAX algorithm for time series anomaly discovery. Section 4 reports the experiments to compare the performance of the two comparative methods on seven time series datasets in two aspects: accuracy and time efficiency. Finally, Sect. 5 gives some conclusions and notes for future studies.

2 Background

2.1 Some Definitions

A time series is a sequence of real numbers measured at equal time intervals. A time series can be a sequence of observations collected from one source, for example, one sensor. In this case, the series is univariate. If we collect information from more than one source, we have a multivariate time series. In this paper, we consider only univariate time series.

Definition 2.1. *Non-self match*: Given a time series T including a subsequence C of length n starting from position p and a matching subsequence M starting from the position q. If $|p - q| \geq n$, M is called as a non-self match to C.

Definition 2.2. Time series 1-discord: Given a time series T, containing the subsequence C of length n beginning at position p. if C has the largest distance to its nearest non-self match, C is called as the top discord (1-discord) of T.

The 1-discord in a time series is also called the top anomaly subsequence in that time series.

2.2 Taxonomy of Time Series Anomaly Detection Methods

The methods for time series anomaly detection are grouped into four categories: window-based methods, segmentation-based methods, prediction-based methods and classification-based methods.

Window-based method extracts fixed length windows (subsequences) from the time series and computes the distances between the current subsequence with all other subsequences in the time series (using some distance measure). The subsequence which has the largest distance to its closest matching subsequence is considered as the top anomaly pattern. HOT SAX is a typical algorithm for finding the top anomaly in time series which belongs to window-based category. The top anomaly detection algorithm can be extended to become the top-k anomaly detection algorithm.

Prediction-based method uses a prediction model to fit the time series and obtains the predicted value using on the past data. Points (subsequences) that deviate remarkably from their predicted values are determined as anomaly points (subsequences). The predictor used in this anomaly detection approach can be a statistical models such as Auto Regression (AR), ARMA, ARIMA or a machine learning model such as artificial neural network, Support Vector Regression (SVR).

Classification-based method uses some technique to extract fixed length subsequences from the training part of a time series and labels each of them as normal pattern or anomaly pattern. Using the set of these class labeled patterns, a classifier is trained and it can be used to classify a new subsequence as normal or anomaly pattern.

Segmentation-based method uses some segmentation technique to split a time series into segments (subsequences). Then a clustering algorithm is used to group the subsequences into clusters. Using the results of clustering, anomaly scores of all the subsequences will be determined and the subsequences of which the anomaly scores are higher than a given threshold will be considered as anomaly patterns.

Prediction-based methods and classification-based methods can be viewed as supervised or semi-supervised anomaly discovery methods while window-based methods and segmentation-based methods can be viewed as unsupervised anomaly discovery methods.

The experimental results in the survey on time series anomaly detection by Chandola et al., in 2009 [16], revealed that generally, window-based methods tend to outperform prediction-based methods.

2.3 Long Short Term Memory

Long Short-Term Memory (LSTM) [17] is an improved variant of Recurrent Neural Network (RNN) designed specifically for sequential data. Each LSTM unit is a generalization of RNN unit, such that part of information about previous time series data points is stored into the network.

Each LSTM unit has three gates:

- Forget gate, which is responsible for deciding which part of information from the previous state should be saved or thrown away.
- Output gate, which is responsible for selecting how much information should be output.
- Input gate, which is responsible for obtaining new information.

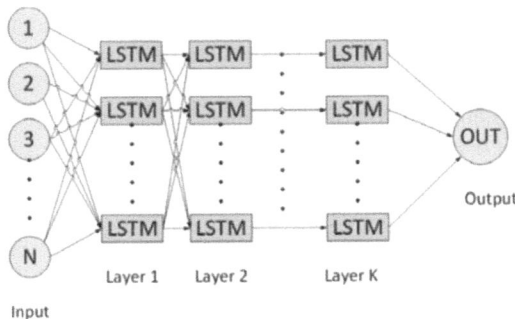

Fig. 1. Stacked Long Short Term Memory network

The deep LSTM neural network includes more than one hidden layer. It consists of multiple layers of LSTM units in which the outputs of the previous layer become the inputs of the next layer. Figure 1 describes the structure of a stacked Long Short Term Memory network which can be used in time series prediction.

3 The Selected Comparative Methods for Time Series Anomaly Detection

In this section, we describe the LSTM-based anomaly detection method and HOT SAX algorithm.

3.1 A Prediction-Based Method: LSTM-Based Anomaly Detection

In this study, we use LSTM-based method for time series anomaly detection which belongs to the category of prediction-based methods. Previous study shows that LSTM-based method is a strong baseline for time series prediction, which outperforms several other prediction methods. Due to this reason, stacked LSTM is a good choice to be predictor in a prediction-based method for anomaly detection. As for time series anomaly detection, inspired by the LSTM_AD method (Malhotra et al., 2015 [12]), we model the normal behavior of a time series through a stacked LSTM network and detect deviations from normal behavior as anomaly patterns. For the prediction task in anomaly detection, we apply multistep-ahead prediction strategy. The workflow of LSTM-based method for time series anomaly discovery is described as the two following steps.

Prediction task
We first train a prediction model using stacked LSTM network and then compute the prediction error distribution using which we detect anomalies. The prediction model accepts the input consisting of p data points in the most recent past and predicts the output consisting of q future values. We stack LSTM layers such that each unit in a lower LSTM hidden layer is fully connected to each unit in the LSTM hidden layer above it through feedforward connections (see Fig. 1). There is one unit at the output layer (also called dense layer). Here we apply multistep-ahead prediction with *recursive strategy* [18]. We use linear transfer function at the input units and MSE (mean squared error) as error function. The model is trained on normal part of the time series in order that it can learn the normal behavior of the time series.

Through multistep-ahead prediction with *lookahead* parameter q at the time point t, the model will predict q values in the future (i.e. at $t + 1, t + 2, ..., t + q$).

Detecting anomaly patterns
The anomaly detection can be done by using prediction errors. Prediction error means the deviation between the predicted value and the actual value at time point t. The prediction errors can be modeled by using Gaussian distribution function. The parameters of Gaussian function: the mean μ and the standard deviation ρ can be estimated from the prediction errors using Maximum Likelihood Estimation (MLE). Probability density (PD) function is defined by:

$$P(x) = \frac{1}{\sigma\sqrt{2\pi}}e^{-(x-\mu)^2/2\sigma^2} \tag{1}$$

As for prediction errors, to avoid vanishing problem when the values becoming so small, the logarithms of probability density values (log*PD*) of prediction errors should be computed and used as a measure to detect anomalies. A test part of dataset which contains normal data and abnormal data is used to determine the *threshold* log*PD value*. This threshold value will be used to detect anomaly values. In the test part, any data point of which the log*PD* value is smaller that the threshold log*PD* will be considered as anomaly data point.

The algorithm

From the two above-mentioned steps, we come to the LSTM-based algorithm for anomaly detection in time series as follows.

Each given time series is divided into four parts. N: the training set which contains only normal data points will be used to train the LSTM model. V_N: the validation set which contains only normal data points will be used to evaluate the model in training stage. V_A: the validation set which contains both normal data points and abnormal data points will be used to compute the threshold for log*PD*. T: the test set which contains both normal data points and abnormal data points will be used to detect anomaly patterns in the time series. The steps of the algorithm are described as follows:

Step 1: The training set N is used to train the LSTM model. All the hyper-parameters of this model is determined through experiments.

Step 2: The validation set V_A is used in the model training stage to early stop the training process and to alleviate overfitting.

Step 3: The validation set V_N is used to collect the differences between the predicted values and the corresponding actual values and use MLE to determine the mean μ and the standard deviation ρ for the Gaussian distribution function.

Step 4: The log*PD* values for all prediction errors in the validation set V_A are computed and based on these values, the threshold value for log*PD* in detecting anomalies is determined.

Step 5: Based on the threshold log*PD* value, the process of finding anomalies will be performed on the test set T.

3.2   A Window-Based Detection Method: HOT SAX

HOT SAX, proposed by Keogh et al., 2005 [1], is a time series anomaly detection algorithm which belongs to window-based category. HOT SAX applies PAA (Piecewise Aggregate Approximation) for dimensionality reduction [19], uses Symbolic Aggregate Approximation (SAX) (Lin et al., 2003 [20]) as a discretization transform and utilizes two ordering heuristics for the inner and outer loops to improve the top anomaly search process.

One interesting property of HOT SAX is that there have been several improved variants of HOT SAX, which brings out remarkably higher time efficiency. Some improved variants of HOT SAX can be reviewed as follows. Buu and Anh in 2011 devised HOTi-SAX which brings out the same accuracy as HOT SAX and executes about 4 times faster than HOT SAX [21]. Thuy et al. in 2016, proposed Hash_DD, an improved version of HOT SAX which yields the same accuracy as HOT SAX and executes about 8.4 times faster than HOT SAX [22].

Besides, Anh and Hien, in 2021 [23] proposed a time series anomaly detection algorithm, called DPDD, which applies dynamic programming into Brute-Force algorithm [1] to bring out the same accuracy as HOT SAX and the computational efficiency about 25.2 times faster than HOT SAX. Thuy et al. in 2018 [24] devised a segmentation-based approach for time series anomaly detection, called EP-ILeader, which yields the same accuracy as HOT SAX and executes about 244 times faster than HOT SAX.

4   Experimental Evaluation

In this section, we describe the experiments for comparing the performance of two methods: HOT SAX and LSTM-based anomaly detection method (abbreviated as LSTM_DD). The two methods were implemented with Python. To implement LSTM_DD we also utilize the open-source framework Keras 2.3.1 with the core Tensor-Flow [25]. The experiments were conducted on Macbook Pro 2020 - MWP42; 2.0 GHz Quad-Core Intel Core i5 gen-10th; RAM: 16 GB.

4.1   Data Description and Parameter Setting

We conducted the experiments over seven time series datasets. Most of the datasets are from the UCR Time Series Data Mining Archive for anomaly detection [26], except the Numenta dataset which is from [27]. The datasets are from different application domains (finance, medicine, industry, image processing). These seven benchmark time series are commonly used by the research community on time series anomaly detection.

Table 1 summarizes the domain and the length of each tested datasets. Figure 2, 3, 4, 5, 6, 7, 8 illustrate the plots of seven tested time series. Notice that Ann_gun dataset is from image processing and therefore is a two-dimensional time series (see Fig. 8).

The anomaly patterns in these benchmark datasets have been *annotated* by experts. For example, TEK16 is a sensor time series representing normal Space Shuttle Marotta Valve Time Series annotated by a NASA engineer [1]. In the plot of TEK16 shown in Fig. 5, NASA engineer annotates the 1-discord (bold in red) of this time series occuring at the fifth cycle of the time series. Another example is the Power-demand dataset which measured the power consumption for a Dutch research facility for the entire year of 1997. In the plot shown in Fig. 4, the three top discords (which are bold in green, red and purple) in Power-demand dataset are annotated by experts. The annotations given by experts in each time series are very useful in checking the accuracy of discord detection by a given method.

The parameters for HOT SAX are the discord length n, the length of PAA frame w and the alphabet size a. To estimate the discord length n and the length of PAA frame w, we applied the segmentation-based techniques which were proposed in the work [28] by Thuy et al. to facilitate parameter determination in HOT SAX. All the parameters for HOT SAX are reported in Table 2. Interested readers on how to determine parameters in HOT SAX can refer to the work [28].

As for the architecture of stacked LSTM in LSTM_DD, for most of time series datasets, it consists of two hidden layers and one dense layer. Paricularly for Power-demand dataset, it is made of one hidden layer and one dense layer. We determine

Table 1. Details of the tested datasets

Name	Description	Length
ECG	Medicine	18000
Numenta	Industry	22695
Power-demand	Industry	35040
TEK 16	Industry	5000
Stock	Finance	5000
Memory	Industry	6875
Ann-Gun	Image processing	11250

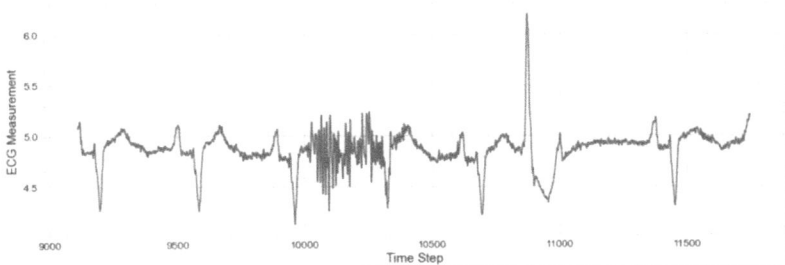

Fig. 2. The ECG time series

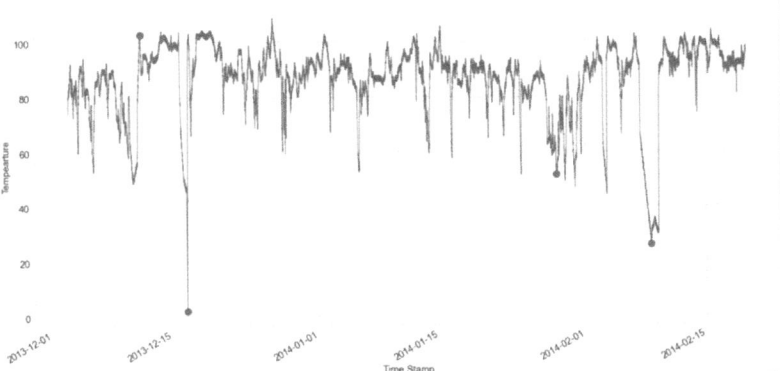

Fig. 3. The Numenta time series

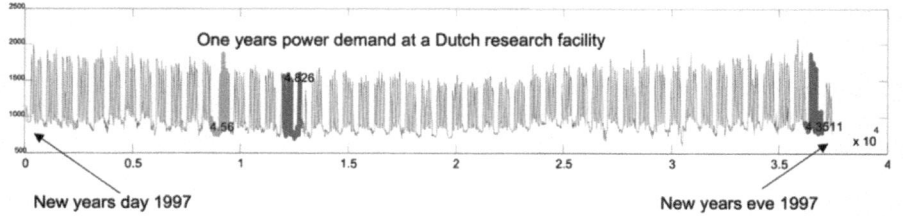

Fig. 4. The Power-demand time series

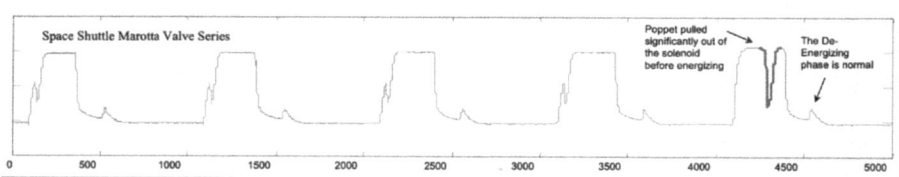

Fig. 5. The TEK16 time series

Fig. 6. The Stock time series

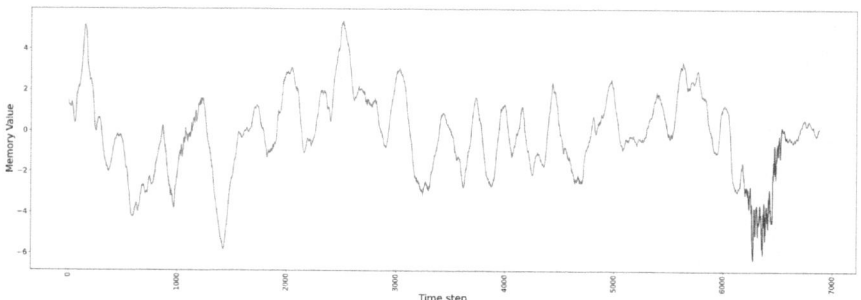

Fig. 7. The Memory time series

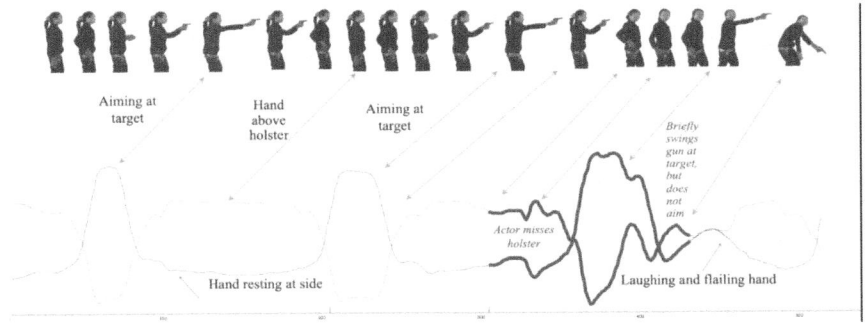

Fig.8. The plots of 2-D Ann_gun dataset

Table 2. Parameters in HOT SAX for seven datasets

Dataset	Discord length	PAA frame size	Alphabet size
ECG	n: 128	w: 8	a: 3
Numenta	n: 128	w: 8	a: 3
Power-demand	n: 1000	w: 8	a: 3
TEK 16	n: 128	w: 8	a: 3
Stock	n: 128	w: 8	a: 3
Memory	n: 128	w: 8	a: 3
Ann-Gun	n: 128	w: 8	a: 3

the hyperparametes of stacked LSTM in LSTM_DD through experiments. The stacked LSTM networks are trained with Adam optimizer [29] in which learning rate (η) and decay factor must be determined. All the parameters for LSTM_DD are given in Table 3. As for the logPD threshold in LSTM_DD for each time series dataset, we determined the suitable values as follows:

ECG: -5; Numenta: -25; Power-demand: -28;
TEK 16: -20, Stock: -5; Memory: -13;
Ann-Gun_X: -8; Ann-Gun_Y: -9.

4.2 Detection Accuracy

The key observations for LSTM-DD method from our experimental results are as follows. In Fig. 10, the logPD values in the anomaly regions are significantly lower than those in normal regions for the ECG dataset. Figure 10 (top) illustrates three plots for ECG time series in LSTM_DD (true value: green, predicted value: dashed green, error: red). Figure 10 (bottom) shows the logPD plot for ECG time series (logPD: green, logPD threshold: dashed green). We can see that this time series dataset has only one anomaly subsequence detected at the data points which have logPD values below the logPD

threshold. This anomaly detection technique used in ECG dataset are repeated in all other datasets.

Table 3. Parameters in LSTM_DD for seven datasets

Dataset	LSTM structure	Adam optimizer	Lookback, lookahead	Batch-size
ECG	1st layer: {60} Dropout: 0.1 2nd layer: {30} Dropout: 0.1 Dense layer: {1}	η (learning rate): 0.1 Decay: 0.99	8, 5	256
Numenta	1st layer: {80} Dropout: 0.1 2nd layer: {20} Dropout: 0.1 Dense layer: {1}	η: 0.05 Decay: 0.99	24, 12	1024
Power-demand	1st layer: {300} Dropout: 0.2 Dense layer: {1} Linear Activation	η: 0.01 Decay: 0.99	1, 1	672
TEK 16	1st layer: {80} Dropout: 0.2 2nd layer: {30} Dropout: 0.2 Dense layer: {1}	η: 0.01 Decay: 0.99	1, 1	1000
Stock	1st layer: {60} Dropout: 0.1 2nd layer: {30} Dropout: 0.1 Dense layer: {1}	η: 0.1 Decay: 0.99	5, 3	256
Memory	1st layer: {60} Dropout: 0.1 2nd layer: {30} Dropout: 0.1 Dense layer: {1}	η: 0.1 Decay: 0.99	5, 3	256
Ann-Gun	1st layer: {80} Dropout: 0.2 2nd layer: {30} Dropout: 0.2 Dense layer: {1}	η: 0.01 Decay: 0.99	1, 1	150

As for checking the effectiveness of the two comparative methods, we compare the resulting results given by the two methods with the expected results from the annotations given by experts based on the domain meaning of the dataset.

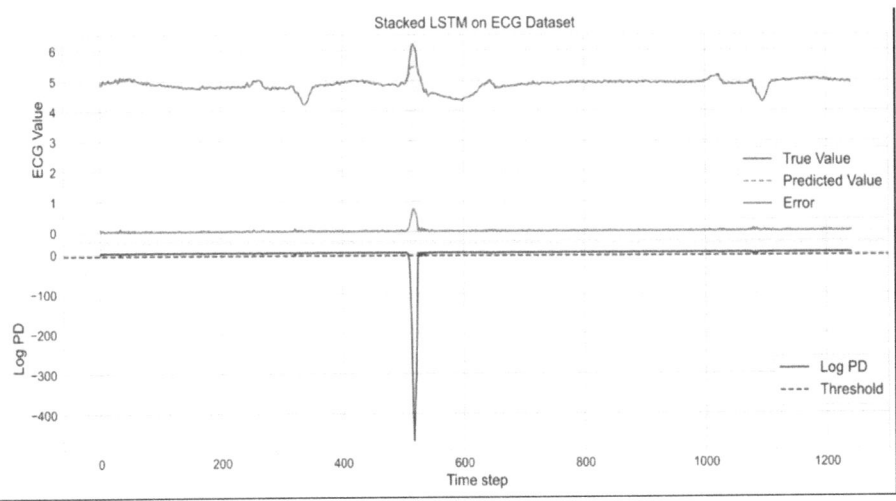

Fig. 10. Illustration of the results found by LSTM-DD on ECG dataset

From our experiments, mainly through inspection by eye, we discovered that the resultant top anomaly subsequence detected by LSTM_DD for each test dataset almost matches with the one discovered by HOT SAX and both of them match with the one annotated by experts for the dataset. Besides, for some datasets, the start location of the top anomaly subsequence detected by LSTM_DD might be slightly different from that of the one found by HOT SAX. However, this kind of differences affects slightly on the detection accuracy.

In sum, it is clear that both LSTM_DD and HOT SAX bring out the same detection accuracy in time series anomaly detection.

4.3 Efficiency

To check the time efficiency of the two comparative methods, we collected the running times (in seconds) of the two methods on seven datasets. Table 4 reports the running times of the two methods. It can be seen that the execution times of LSTM_DD are always higher than that of the HOT SAX in all seven datasets. Noted that here the running time of LSTM_DD on a given time series consists of the time for building the prediction model and the time of using the model to detect anomalies. In average, HOT SAX is about 2.58 times faster than LSTM_DD.

Besides, in comparing LSTM_DD with some improved variants of HOT SAX in time efficiency, we get some following results. As mentioned in Subsect. 3.2, since HOTiSAX is about 4 times faster than HOTSAX, HOTiSAX can run about 10.32 times faster than LSTM_DD. Since Hash_DD is about 8.2 times faster than HOT-SAX, Hash_DD can run about 21.156 times faster than LSTM_DD. Since DPDD is about 25.2 times faster than HOTSAX, DPDD can execute about 65.02 times faster than LSTM_DD. As for the segmentation-based method EP-ILeader, since this algorithm can execute about 244 times faster than HOT SAX, it can run about 629.52 times faster than

LSTM_DD. This special finding suggests that efficiency is the key aspect of LSTM_DD method. LSTM_DD for time series anomaly detection should be improved further in computational efficiency to be comparable to window-based approaches.

Table 4. Time comparison of the two methods (in seconds) on seven datasets

Dataset	LSTM_DD	HOT SAX
ECG	399	30
Numenta	246	39
Power-demand	320	318
TEK 16	9	7
Stock	34	9
Memory	27	11
Gun-X	67	17
Gun-Y	51	15

From the experimental results in two performance metrics: detection accuracy and time efficiency, we can find out that LSTM_DD is not superior to HOT SAX since the accuracy of LSTM_DD is almost the same as that of HOT SAX while LSTM_DD incurs a significantly higher computational overhead. The finding from this study on comparing HOT SAX to LSTM-based method in time series anomaly discovery is somewhat consistent with the finding in the important review by Chandola et al. [12] which remarked that generally, window-based methods tend to outperform prediction-based methods.

Furthermore, determining the hyperparameters in LSTM_DD is much more complicated and challenging than estimating the parameters in HOT SAX algorithm.

5   Conclusion

In this study, we give an exploration for the research question whether LSTM-based models are better than HOT SAX in time series anomaly discovery or not. We compared HOT SAX to LSTM-based method on seven benchmark time series datasets in two perspectives: detection accuracy and time efficiency. The experimental results indicate that LSTM-based method does not outperform HOT SAX since the accuracy of LSTM-based method is almost the same as that of HOT SAX while LSTM-based method incurs a significantly higher computational overhead. Therefore, the success of LSTM-based methods in image processing, natural language processing and time series forecasting is not so easy to be replicated in time series anomaly detection. Old-fashion window-based approaches for anomaly detection in time series, such as HOT SAX and its improved variants are still much more efficient than LSTM-based method.

As for future work, we intend to evaluate the two comparative methods on more time series datasets. Besides, we plan to improve further the effectiveness and time efficiency of LSTM-based method in time series anomaly detection by applying a systematic technique in tuning hyperparameters of LSTM-based predictors [30].

References

1. Keogh, E., Lin, J., Fu, A.: HOT SAX: Efficiently finding the most unusual time series subsequence. In: Proceedings of The Fifth IEEE International Conference on Data mining (ICDM), pp. 226–233, (2005)
2. Bu, Y., Leung, T.W., Fu, A., Keogh, E., Pei, J., Meshkin, S.: WAT: Finding top-K discords in time series database. In: Proceedings of the 2007 SIAM International Conference on Data Mining (SDM' 07), Minneapolis, MN, USA, 26–28 (2007)
3. Li, G., Braysy, O., Jiang, L., Wu, Z., Wang, Y.: Finding time series discord based on bit representation clustering. Knowl. Based- Syst. **54**, 243–254 (2013)
4. Oliveira, A.L.I., Meira, S.R.L.: Detecting novelties in time series through neural networks forecasting with robust confidence intervals. Neurocomputing **70**(1–3), 79–92 (2006)
5. Pena, E.H.M., de Assis, M.V.O.M., Proença Jr., M.L.: Anomaly detection using forecasting methods ARIMA and HWDS, In: Proceedings of 32nd International Conference of Chilean Computer Science Society (SCCC), Temuco, Chile, pp. 11–15 (2013)
6. Yu, Y., Zhu, Y., Li, S., Wan, D.: Time series outlier detection based on sliding window prediction. Math. Problems Eng. **2014**, 879736, (2014)
7. Siami-Namini, S., Tavakoli, N., Sinam-Namin, A.: A comparison of ARIMA and LSTM in forecasting time series, In: Proceedings of 17th IEEE International Conference on Machine Learning and Applications (ICMLA), 17–20, Orlando, FL, USA, pp. 1394–1401 (2018)
8. Sezer, O.B., Gudelek, M.U., Ozbayoglu, A.M.: Financial time series forecasting with deep learning: a systematic literature review: 2005–2019, Appl. Soft Comput. **90**, 106181 (2020)
9. Han, Z., Zhao, J., Leung, H., Ma, K.F., Wang, W.: A review of deep learning models for time series forecasting, IEEE Sens. J. **21**(6), 7833–7848, (2021)
10. Lindermann, B., Muller, T., Vietz, H., Jazdi, N., Weyrich, M.: A survey on long short term memory networks for time series prediction. Procedia CIRP **99**, 650–655 (2021)
11. Lindemann, B., Maschler, B., Sahlab, N. and Weyric, M.: A survey on anomaly detection for technical systems using LSTM networks. Comput. Ind. **131**, 103498 (2021)
12. Malhotra, P., Vig, L., Shroff, G., Agarwal, P.: Long short term memory networks for anomaly detection in time series. In: Proceedings of European Symposium on Artificial Neural Networks (ESANN), Bruges (Belgium), 22–24 April, pp. 89–94 (2015)
13. Chauhan, S. Vig. L.: Anomaly detection in ecg time signals via deep long short-term memory networks. In: Proceedings of 2015 IEEE International Conference on Data Science and Advanced Analytics (DSAA), Paris, France, 19–21 October, pp. 1–7 (2015)
14. Buda, T., Caglayan, B. Assem, H.: Deepad: A generic framework based on deep learning for time series anomaly detection. In: Proceedings of Pacific-Asia Conference on Knowledge Discovery and Data Mining (PAKDD 2018), LNCS 10937, Springer, Cham, pp. 577–588 (2018)
15. Zhang, L., Yang, L., Gu, C. Li, D.: Lstm-based short-term electrical load forecasting and anomaly correction, E3S Web of Conferences **182** (01004) (2020)
16. Chandola, V., Cheboli, D. K., Kumar, V.: Detecting anomalies in a time series database, Technical Report, Department of Computer Science and Engineering, University of Minnesota, TR-09-004 (2009)
17. Hochreiter, S., Schmidhuber, J.: Long Short-Term Memory. Neural Comput. **9**(8), 1735–1780 (1997)
18. Ben Taieb, S., Bontempi, G., Atiya, A. F. Sorjamaa, A.: A review and comparison of strategies for multi-step ahead time series forecasting based on the {NN5} forecasting competition. Expert Syst. Appl. **39**(8), 7067–7083 (2012)
19. Keogh, E., Chakrabatti, K., Pazzani, M.: Dimensionality reduction for fast similarity search in large time series databases. Knowl. Inf. Syst. **3**, 263–286 (2001)

20. Lin, J., Keogh, E., Lonardi, S. Chiu, B.: A symbolic representation of time series, with implications for streaming algorithms, In: Proceedings of the 8th ACM SIGMOD Workshop on Research Issues in Data Mining and Knowledge Discovery, DMKD 2003, pp. 2–11 (2003)
21. Buu, H.T.Q., Anh, D.T.: Time series discord discovery based on iSAX symbolic representation. In: Proceedings of the 3rd International Conference on Knowledge and Systems Engineering (KSE), Hanoi, Vietnam, 14–17 October, pp. 11–18 (2011)
22. Thuy, H.T.T., Anh, D.T., Chau, V.T.N.: An effective and efficient hash-based algorithm for time series discord discovery, In: Proceedings of the 2nd National Foundation for Science and Technology Development Conference on Information and Computer Science (NICS 2016), September 14–16, Da Nang, Vietnam, pp. 85–90 (2016)
23. Anh, D.T. Hien, N.V.: A dynamic programming approach for time series discord detection, In: Proceedings of 10th EAI International Conference on Context-Aware Systems and Applications (ICCASA 2021), Virtual Event, 28-Oct, pp. 255–266 (2021)
24. Thuy, H.T.T., Anh, D.T., Chau, V.T.N.: Novel method for time series anomaly detection based on segmentation and clustering. In: Proceedings of 10th International Conference on Knowledge and System Engineering (KSE), IEEE, Ho Chi Minh City, Vietnam, 1–3 November, pp. 276–281 (2018)
25. Cholett, F.: Keras. http://keras.io. Accessed 2021
26. The UCR Time Series Dataset Archive for Discord Detection http:/www.cs.ucr.edu/~eamonn/discords/. Accessed 2021
27. Lavin, A., Ahmad, S.: Evaluating real-time anomaly detection algorithms: the numenta anomaly benchmark. In: Proceedings of IEEE 14th International Conference on Machine Learning and Applications (ICMLA), Miami, Florida, USA, 2–11 December (2015)
28. Thuy, H.T.T., Anh, D.T., Chau, V.T.N.: Some segmentation-based techniques to improve time series discord discovery, In: Proceedings of International Conference on Nature of Computation and Communication (ICCTC 2016), March 17–18, Rach Gia, Vietnam, LNICST 128, Springer, pp. 179–188 (2016)
29. Kingma, D. B., Ba, J.: Adam: a method for stochastic optimization, arXiv preprint arXiv :14126 980 (2014)
30. Abbasimehr, H., Shabani, M., Yousefi, M.: An optimized model using LSTM network for demand forecasting. Comput. Ind. Eng. **143**, 106435 (2020)

Context-Aware Technologies

Application of Machine Learning Models for Predicting Glucose-Level in the Pure Fluid with Algorithm for Reducing Data Dimension Based on Data Series Extraction

Tri Ngo Quang, Tung Nguyen Thanh[✉], Huong Pham Thi Viet, and Huy Bui Quang

International School, Vietnam National University, Hanoi, Vietnam
{tung_nt,huongpv}@vnu.edu.vn

Abstract. The phenomenon that glucose level of pure liquid is able to define patterns of Raman spectroscopy was demonstrated in several studies. Nevertheless, it is difficult to predict glucose level accurately by manual methods so machine learning techniques are proposed to support it. In the range of the report, we employ three simple machine learning models including Extra Trees, Random Forest, and SVM to predict glucose level from Raman spectroscopy of pure water-mixed fluid which we collected by infrastructures of Vietnam National University. In addition, the Raman data was simplified by dimension reduction algorithms based on handling data series. The results show the effectiveness of the machine learning models for predicting glucose levels as well as the reduction dimension algorithms for enhancing the performance of machine learning techniques.

Keywords: Raman spectroscopy · machine learning · Diabetes

1 Introduction

The traditional method for diagnosing diabetes is an invasive technique which causes tiny injuries as well as uncomfortable experiences to the patients. Thus, several solutions tend to predict glucose level without drawing blood of patients by implanted lancet and some of them apply the close relation between glucose level and Raman spectroscopy from the reflection in human skin. Numerous studies showed the high viability of non-invasive blood glucose monitoring by Raman waveshift, but the main problem was the challenging data collecting and insufficient precision in calculating blood glucose levels from Raman spectroscopy data. We build on earlier research to suggest a method for predicting glucose level from Raman data in the context of artificial intelligence using machine learning techniques. As a result, we view the measurement as a classification issue involving different glucose-level labels. Our objective in the context of this study is to evaluate the effects of various data pre-processing techniques and validate the potential for glucose-level prediction from Raman waveshifts using machine learning techniques. We have gathered standard data with a high degree of purity and little noise by measuring

© ICST Institute for Computer Sciences, Social Informatics and Telecommunications Engineering 2024
Published by Springer Nature Switzerland AG 2024. All Rights Reserved
P. Cong Vinh and N. Thanh Tung (Eds.): ICCASA 2023, LNICST 579, pp. 87–101, 2024.
https://doi.org/10.1007/978-3-031-58878-5_6

the Raman spectrum of deionized water with a clear glucose level. Additionally, the moderate complexity and broad coverage of the dataset we have gathered make it ideal for the validation objectives of fundamental machine learning models. In order to reduce the complexity of Raman datasets, we also suggest a novel preprocessing technique. In this study, we examined the data's format and put into practice a hotspot series-based data reduction technique. The outcome is very encouraging and shows how machine learning can be used to predict glucose levels using Raman waveshift in addition to showing the benefits of various data pre-processing techniques.

The paper consists of four major sections: theoretical background, technique for analyzing and pre-processing Raman datasets, experiments, and conclusion.

2 Theoretical Background

In this chapter, we describe some general theoretical findings about our project, including non-invasive glucose-level measurement using the Raman spectroscopy and machine learning.

2.1 Non-invasive Glucose Measurement Using Raman Spectroscopy

Raman measurement used the Raman effect, which was discovered by Raman and Krishnan in 1928 [1]. Raman spectroscopy, based on this effect, is a scattering technique. When a sample is exposed to monochromatic laser sources, the molecules in the sample interact with the lasers and scatter light. A Raman spectrum is produced by scattering light at a frequency different from inelastic scattering. The inelastic interaction of sample molecules with monochromatic light generates the Raman spectrum. The measuring instrument that uses Raman spectroscopy is known as a Raman spectrometer and consists of four modules: a laser generator, a chamber for storing measuring samples, a grating-equipped spectrometer chamber, and a detecting system [2].

Before the use of Raman spectroscopy in healthcare was found to be effective, the measurement acquired by non-invasive methods was of concern. For example, Caduff et al. proposed a non-invasive method to predict glucose level from impedance spectroscopy [18]. With the relationship between Raman data and glucose-level, there were several studies that raised huge concerns in the scientific community. Xu et al. [3] described available targets of measurement using Raman spectroscopy as a large amount of substances. Thus, the relationship between Raman spectroscopy reflected from human bodies and blood glucose-level was discussed in [4]. According to this study, blood glucose levels in living animal bodies can be determined in vivo using the Raman spectra obtained from a diode-laser operating at 785 nm. By using partial least squares techniques, the glucose-level of humans can be measured using near-infrared Raman spectroscopy [17]. The research in [19] had a similar approach, but the statistical techniques were classical least squares, principal component regression and partial least squares.

These investigations demonstrate that it is entirely feasible to forecast the glucose level using Raman spectroscopy. Due to the significant difficulty in observing the relationship between Raman data and glucose level, the problem we had to solve was the selection of a prediction technique. We may also notice that several statistical methods have demonstrated their effectiveness in identifying characteristics between Raman spectroscopy and glucose levels. According to previous research and our own findings, the machine learning technique is just as effective as the methods mentioned above. The possibility and approach of machine learning algorithms for glucose prediction based on Raman spectroscopy were addressed in the following section.

2.2 The Use of Machine Learning for Predicting Glucose-Level from Raman Spectra

The Potential of Machine Learning in Glucose Concentration Prediction.
Machine learning has proved its efficiency in problems of prediction and classification [9]. There were several former studies about the application of machine learning on glucose-level classification from extracting Raman data. Some of them concentrated on detecting diabetes from this kind of data. The research in [10] proposed a non-invasive glucose measurement using 5 simple machine learning algorithms from optical sensors. In the study, the dataset was the 18 pairs which each pair had 2 values: wavelength and intensity. The techniques for machine learning included a Feedforward Neural Network for multi-class classification. In this investigation, we investigated the experimental methodology employed by the authors when they collected glucose-distilled water samples to simulate human blood. The objective of this method is to generate a simple dataset with a high degree of accuracy and a predicted value.

In article [16], the author collected Raman spectroscopy data on blood samples from normal people and diabetes patients before proposing an algorithm based on a combination between Principal Component Analysis and Linear Discriminant Analysis to classify [16]. The advantage of the study, when compared to [10] is that Raman spectroscopy was the sole input data to be classified by the algorithms. As reported by [21], the results and classification discretization framework of an experiment using a visible near-infrared laser derived from an optical sensor to measure glucose level were presented and shown to be promising. However, Raman spectroscopy was performed on human blood, which had a high degree of accuracy despite the intrusive nature of blood collection. In addition, 76 individuals were sampled, including 39 diabetes patients and 37 healthy individuals. The small quantity of this number reduces its persuasiveness. The research in [11] introduced a portable spectrometer for non-invasive glucose testing based on the Raman effect. Positive aspect of the study is the in vivo studies conducted on diabetes patients and healthy individuals. Using the spectrometer, the solution acquired Raman spectroscopy from certain regions of the human body via a non-invasive measurement. Support Vector Machine and Artificial Neural Network were used as machine learning prediction models. Some other benefits of the study were the data processing techniques that enhanced the performance of machine learning algorithms. The first problem with this study is the small sample size, which consists of just about 20 individuals. The second categorization is lean, for which it is hard to specify a precise glucose level. Instead, the answer predicted whether or not a sample was obtained from a diabetic patient.

We acquired a number of outstanding perspectives and approaches from the aforementioned studies, while avoiding their flaws. We continue to reference these research while focusing on the outcome and measuring metric. In this section, we present simply the approach and theoretical foundation.

Applied Machine Learning Model.

In the scope of study, we use 3 machine learning models for classification, including Extra Trees, Random Forest and Support Vector Machine (SVM):

- The Extra Trees model:

The Extra Trees machine learning model was proposed based on tree architecture [5]. The Extra-Trees approach follows the traditional top-down method and employs a collection of unpruned decision or regression trees. The number of characteristics that are randomly chosen at each node and the minimal sample size for splitting a node are the two parameters for the Extra-Trees splitting technique for numerical attributes. It is applied several times to the whole original learning sample in order to produce an ensemble model, which we identify by the number of trees in the ensemble. By majority vote in classification problems and arithmetic average in regression problems, the predictions of the trees are combined to get the final prediction.

The Extra Trees model has many applications in detection and classification. A solution for detecting phishing websites integrated Extra Trees algorithms and AI Meta-Learners techniques have been proposed in [6]. Another application of Extra Trees is the Autotrophic/Heterotrophic Microorganism Mixtures using absorbance spectrum data [7].

- The Random Forest model:

The Random Forest model was created by integrating many tree-based data structures and randomizing the dataset and was found to be an effective classifier for bio-sensor data [8]. A number of tree classifiers were integrated with Random Forest based on combination regulation. Each of these classifiers casts a vote for the class with the most members, and the final sort result is produced by combining these votes. This algorithm is distinguished by its high classification precision, tolerance for noise and outliers, and lack of overfitting. During the creation of Random Forest, the tree is also planted on the new training set using random feature selection, and the new training set is taken from the previous training set using bagging methods.

Similar to other methods for machine learning, Random Forest can be used for classification and regression. In [12], several uses of Random Forest for managing complex data from remote sensors are discussed. From the list of these applications, we determined that Random Forest was capable of dealing with a number of data sources, such as multi-spectral radar, sensing images, and hyperspectral imagery. Using Raman spectroscopic data and Random Forests has another use: locating substances used in nanofabrication [13]. The computationally feasible analysis of genome-wide association data using Random Forest is one of the first instances presented in this work.

- The Support Vector Machine model:

The Support Vector Machine (SVM) has a huge variety of supervised applications with different data sources [14]. The capacity of the SVM to learn data classification patterns with a balance between accuracy and reproducibility is what gives it its power. It has gained popularity as a classification tool, though it is still infrequently employed for regression tasks. It is highly versatile and may be utilized in a variety of data science contexts, including the study of brain illnesses. In order for the SVM to function, a hyperplane that optimizes the separation between the support vectors of the two class labels was selected. In comparison to other kinds of classifiers, the SVM's capability and attraction stem mostly from its ability to give balanced performance even when the complexity of the feature space greatly exceeds the number of training data. In addition, the SVM provides diversity. For the SVM decision functions, many distinct kernel functions can be provided, and most software enables users to choose unique kernels. This functionality makes it easier to employ the SVM classifiers to solve linear classification problems without having to spend a lot of time on hyperparameter adjustment. The SVM is effective in solving a variety of classification issues with high dimensional data [20]. Therefore, it can be applied efficiently to Raman spectra datasets.

Both three machine learning models, including Extra Trees, Random Forest and SVM were fully supported in Sci-kit Learn – a Python library for ML implementation. Thus, we use this for developing our project.

3   Collection and Dimensions Reduction of Raman Dataset

3.1   Collection of Dataset

To evaluate the accuracy of machine learning models in predicting glucose levels for the objectives of this study, sample-level precision was crucial. According to prior study, there are three drawbacks to the samples collected through non-invasive or invasive measures in regions of the human body [10]. The dense appearance of noise, the difficulty of obtaining a sample with the desired value, and the medical ethics surrounding human experimentation. We produce a pseudo-sample with the desired value based on these challenges.

With the purpose of replicating blood samples with a determined glucose level in our laboratory, we created samples by combining pure glucose and deionized water in a particular ratio. Main components used for measurements consist of:

- Raman spectrophotometer: uRaman - Ci, Technospex.
- Glucose chemical products: 99.5% Sigma-Aldrich glucose.
- Deionized water solution bottle.
- Tools (solution tube, stirring rod,…).

The measurement procedure is described as below:

1. Set excitation laser power to 100 mW.
2. Set the measurement range at the wavelength of 785.1 nm in 300 s.
3. Place 3 ml of glucose solution from each of the mentioned test tubes on the quartz surface of the Cuvette.
4. Insert the cuvette (with the MACRO-CH/Quartz Cuvette accessory) into the measurement chamber of the uRaman – Ci spectrophotometer.
5. Collect digitized Raman signals from the measurement experiment.

The set of devices used to acquire Raman spectroscopic data by analyzing samples of glucose-mixed fluids with Technospex uRaman - Ci. We have not yet adopted non-invasive measurement in human tissue due to the side effects of external variables in the Raman data from this measurement were too significant to permit an evaluation of the correlation between glucose level and sample figures.

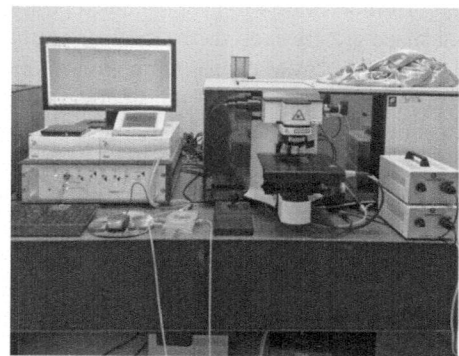

Fig. 1. Device set used for Raman spectroscopy measurement from artificial glucose-mixed fluids.

Within the scope of the study, our dataset contains 50 samples with 50 labels, but only 10 label-values. The Raman shift data is contained in a CSV file, and its label is the prefix of this CSV file, which is separated from the other portions by a "-" character. Each label in the set indicates a glucose value and was encoded as an integer: $I = \{1, 2, 3, 4, 5, 6, 7, 8, 9, 10\}$. The encoding process was handled by a function which analyzed the CSV file. Basic description about our dataset is described in Table 1:

Table 1 Description of the dataset used for the machine learning algorithm.

Glucose concentration level (mmol/L)	Number of samples	Label code
5.0	5	1
5.5	5	2
6.0	5	3
6.5	5	4
7.0	5	5
7.5	5	6
10.0	5	7
12.5	5	8
15.0	5	9
20.0	5	10

The distance between values indicated by labels is diverse, as seen in Table 1. Between 15.0 and 20.0, the shortest distance is 0.5 m and the longest is 5.0 m. The gap between glucose levels is not considered as an extraction criterion because the machine learning technique was built to handle only classification problems. Yet, under conventional settings for measuring blood samples, glucose concentration is the most influential element in determining the result pattern. If the difference in glucose levels between two samples was modest, there would be a high degree of data pattern similarity, which would make machine learning algorithm extraction more challenging. Consequently, the dataset enables us to test the performance of several machine learning models for extracting features from different glucose feature distances.

3.2 The Hotspot Series Extraction Procedure

The Primitive Input Data.
Additionally, the moderate complexity and broad coverage of the dataset we have gathered make it ideal for the validation objectives of fundamental machine learning models. In order to reduce the complexity of Raman datasets, we also suggest a novel preprocessing technique. In this study, we examined the data's format and put into practice a hotspot series-based data reduction technique. The outcome is very encouraging and shows how machine learning can be used to predict glucose levels using Raman waveshift in addition to showing the benefits of various data pre-processing techniques. Based on this explanation, we define each sample's input data as an array of intensity values, where each intensity value's index corresponds to its index within the sample. The fundamental input data for the machine learning method is specified as a 2048-element array of intensity values sorted ascendingly by matching waveshift throughout the scope of the study.

Hotspot Segments of Primitive Data.
We recognize that the primitive input data of a sample is extremely complicated and needs to be reduced. In each labeled-group, we randomly selected one of five samples and plotted it using the Python tool matplotlib, as shown in Fig. 2.

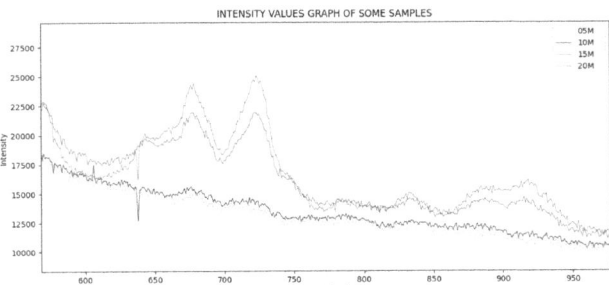

Fig. 2a Large distance between glucose levels of samples: 5, 10, 15 and 20

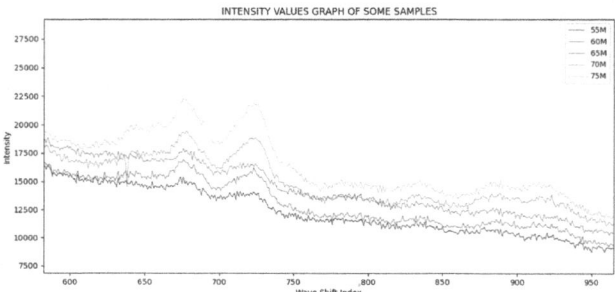

Fig. 2b Small distance between glucose levels of samples: 5.5, 6.0, 6.5, 7.0 and 7.5 (The graph of intensity values in some samples with different glucose level)

Figure 2 contains two plots with distinct glucose level separations. In Fig. 1a, the distance is 5, and the range of glucose levels is from 5 to 20. In contrast, the distance in Fig. 1b is only 0.5, and the range of glucose levels is between 5.5 and 7.5. We can observe that:

- In all samples, the ascending sequence of wave shift does not provide a persistent change in intensity. There are a considerable number of 10-value curves with the maximums in the initial 1250 indexes. Throughout the last 750 indices, the number has fluctuated modestly.
- In the first indexes of these graphs, the difference between samples' glucose concentrations is expressed clearly. In the range of indexes from 500 to 1250, visibility is excellent.

This finding leads us to the conclusion that the length of input data is enormous, but not all values provide useful classification information for machine learning models. Hence, we define a hotspot as a chunk of input data containing useful categorization features. By examining the values of the smaller-than-input-data hotspot region, machine learning models may collect nearly all the characteristics of a sample with a certain glucose level. Although we can define hotspot segments through observation, we build a method to detect them with greater precision.

Generation of Intensity Series.
First, we define an intensity series as a portion of input data containing a predetermined number of sequence values. As mentioned previously, each sample's input data is an intensity array. This array can be represented in the following mathematical formula (1):

$$A_i(g) = \{I_j \in N \,|\, j \in [0, 2047]\} \tag{1}$$

In the formula (1):

- $A_i(g)$: array of intensity values with index i and glucose level g
- I_j: intensity value with index j

This array is used to construct an intensity series of length l (l < 2048) by selecting l sequence elements. Hence, the distinct series are produced by assigning distinct elements to the series' initial constituents. The generic formula (2) provides the mathematical structure of a length l intensity series:

$$S_j(A_i(g)) = \{I_k \in N \,|\, l \in [1, 2047],$$

$$j \in [0, 2048 - l], k \in [j, j + l]\} \tag{2}$$

In the formula (2):

- j: starting index of an intensity series
- $A_i(g)$: array of intensity values with index i and glucose level g
- $S_j(A_i(g))$: an intensity series of $A_i(g)$ with starting index j
- I_k: intensity value with index k. This index is setted in $A_i(g)$.
- l: the size of the intensity series

Because the intensity array has 2048 indexes, each sample has a set of (2049 - l) series with a separate starting index. From this set, we select the hotspot series and generate a new array that contains all elements of the hotspot series.

Extraction of Hotspot Series.
After extracting the intensity series, the next step is selecting the hotspot series and defining the new input data with similar extraction of all samples. Thus, the extraction process includes 3 stages: calculate hotspot level, select the hotspot series based on the hotspot level and output new input data for the machine learning models.

In the first stage, we calculate hotspot level based on median and average variance of this series in the Formula (3) and (4):

$$\underline{S}_j(A_i(g)) = \frac{1}{l} \sum_{j+l}^{j} (I_j) \tag{3}$$

In the formula (3):

- $A_i(g)$: array of intensity values with index i and glucose level g
- $\underline{S}_j(A_i(g))$: a median of the intensity series of $A_i(g)$ with starting index j
- I_j: intensity value with index j
- l: the size of the intensity series

These is formula (4)

$$V_j(A_i(g)) = \frac{1}{l} \sum_{j+l}^{j} \left(I_j - \underline{S}_j(A_i(g))\right)^2 \tag{4}$$

In the formula (4):

- $A_i(g)$: array of intensity values with index i and glucose level g
- $V_j(A_i(g))$: average variance of the intensity series of $A_i(g)$ with starting index j
- $\underline{S}_j(A_i(g))$: a median of the intensity series of $A_i(g)$ with starting index j
- I_j: intensity value with index j
- l: the size of the intensity series

After calculating the average variance of each array's series, we chose a set number of the series with the highest average variance. Then, new array input data are defined by picking all indexes of the intensity array that are contained in the selected series of all samples. The new input data is the intensity array from which all intensity values whose indices are not selected have been removed.

4 Experiment

4.1 Setup of Experiments

We implemented our machine learning algorithm with three models, including Extra Trees, Random Forest and Support Vector Machine on a personal computer with adequate software and hardware configuration. Our computer has 8GB of RAM capability and a Microsoft 64-bit operating system. Meanwhile, Python was selected to implement this algorithm because this programming language has a large number of libraries that strongly support machine learning models. Specifically, we use Python-based libraries such as Scikit-learn to construct these machine learning models as well as Mathplot-lib to write graph about Raman spectra in Fig. 2.

4.2 Creation of Experimental Dataset

Implementation of the Hotspot Series Extraction Algorithm.
We implement the hotspot series extraction algorithm using Python. We set the size of the series as 10 values and the number of selected series in each sample is 500 series. With this dataset, the result is that 831 values are selected with some segments including index from 0 to 29, index from 31 to 542, index from 553 to 586, index from 629 to 647, index from 664 to 699, index from 703 to 743, index from 782 to 799, index from 992 to 1010, index from 1109 to 1122, index from 1171 to 1207, index from 1505 to 1523, index from 1620 to 1642, index from 1706 to 1715 and index from 1865 to 1883. In conclusion, the new input data has 831 values which contain 40.5% in comparison with the size of primitive data.

The next step is to divide this experimental dataset into a train set and a test set. We subdivided the dataset based on k-fold validation to include data for training and testing.

Separation of Dataset.
The next step is to divide this experimental dataset into a train set and a test set. We subdivided the dataset based on k-fold validation to include data for training and testing.

To avoid overfitting, we implement k-fold cross validation, which is the process of using each subset as the test data set and the remaining subsets as the training data. It involves breaking a data set into k subsets. The performance metrics for each validation process are then averaged. There is not a single best indicator for evaluating machine learning algorithms because each approach has advantages and disadvantages. In the experiment, we divided the population equally into 5 subsets, with one subset used for the test set and the other four subsets used for the training set.

The test set for root dataset consists of 10 samples, each of which contains a distinct label-value from a collection of 10 label-codes. Currently, the train value data consists of forty samples, four of which have identical label-codes. The test set comprises six samples, each of which has a unique label-value selected from a collection of six label-codes. In the interim, the train value data includes 24 samples, four of which have identical label codes.

In addition, there are 450 iterations of training with each machine learning model including Extra Trees, Random Forest and SVM models.

4.3 Measuring Criterions

The algorithm's efficiency is determined by the accuracy of the classification process with the data from the test set. Specifically, the algorithm with a specific machine learning model classifies a sample regardless of its label-code and defines a label-code for this sample. After that, the algorithm compares the predicted label-code to the existing label-code of this sample. There are possible 2 cases of this comparison:

- *True (T):* The predicted label-code and existing label-code is the same.
- *False (F):* The predicted label-code and existing label-code is different.

There are 4 criterions used to determine the effectiveness of our model including: Accuracy (Acc), Specificity (Sp), Sensitivity (Se) and ROC-AUC of our model. Meanwhile, Specitivity, Sensitivity and ROC-AUC are calculated by One-vs-rest (OvR) strategy:

The accuracy of our model is defined by Formula (5):

$$Acc = \frac{TP + TN}{TP + FP + TN + FN}, \tag{5}$$

The model's Specificity (Sp) is defined as:

$$Sp = \frac{TN}{TN + FP} \tag{6}$$

The Sensitivity (Se) is defined as.

$$Se = \frac{TP}{TP + FN}, \tag{7}$$

where

- TP – Number of true positive instances
- TN – Number of true negative instances
- FP – Number of false positive instances
- FN – Number of falsenegative instances

Each machine learning model consists of $i = 450$ loops in which subsets are differently partitioned into train and test sets prior to training and accuracy values are determined after each turn. Afterwards, the mean metrics accumulated over 450 iterations are calculated. There are 3 ML model used in our scope of investigation: Extra Trees, Random Forest, and Support Vector Machine; thus, there are three values of average Accuracy for each model.

4.4 Results of Experiments

This section discusses the outcome of applying our selected augmentation strategies for Raman spectroscopy to several ML models. The following are 3 scenarios for the dataset:

- Root dataset - without using any preprocessing method
- Hotspot series dataset – Using hotspot series extraction applied to the root dataset.

Root Dataset - Without using Hotspot Series Extraction.
Table 2 shows the result with each model in case of primitive data without any conversion. In each sample, the number of intensity figures is 2048. The metrics are described in percentage unit (%):

Table 2 Result of experiment with each model in case of root dataset.

	Mean result over 450 iterations			
	Acc	Sp	Se	ROC-AUC
Extra Trees	92.64	99.19	92.73	99.23
Random Forest	87.24	87.47	85.40	98.59
Support Vector Machine	84.43	98.38	86.21	99.04

The Extra Trees model has the highest accuracy score in this table, whereas the Random Forest and SVM models have lower Average Accuracy values. The accuracy ratings range from 84% to 92%, which is a moderately good range. Similar to the accuracy scores, the Extra Trees model has greater Specificity, Sensitivity, and ROC-AUC values than the other models. While the SVM model's estimate of its learning capacity is superior, its performance with this data is the worst. Yet, when comparing one component of One-vs.-Rest Specificity to another, we detect an anomalous characteristic. The SVM model has a better specificity than Random Forest, which is distinct from the three other measures. It indicates that the SVM model is more resistant to type I errors due to its unique processes for tackling classification issues.

Preprocessed Data - Using Hotspot Series Extraction Algorithm.
Table 4 shows the result with each model in case of data after using hotspot series extraction. In each sample, the number of intensity figures is only 831 which means its size is 40.5% of the size of the root dataset. The basic unit is also percent (%):

Because accuracy is the most crucial criterion, we compare value between different models as well as between root and preprocessed dataset and describe visually in the Fig. 3:

From Table 3 and Fig. 6, the Extra Trees model has the highest accuracy score in this table, whereas the Random Forest and SVM models have lower Average Accuracy values. The accuracy ratings range from 84% to 92%, which is a somewhat good range. Similar to the accuracy scores, the Extra Trees model has greater Specificity, Sensitivity,

Table 3 Result of experiment with each model in case of hotspot series preprocessed dataset

| | Mean result over 450 iterations | | | |
	Acc	Sp	Se	ROC-AUC
Extra Trees	93.77	99.30	93.67	99.48
Random Forest	88.24	98.70	99.30	99.02
Support Vector Machine	89.63	98.84	98.60	99.37

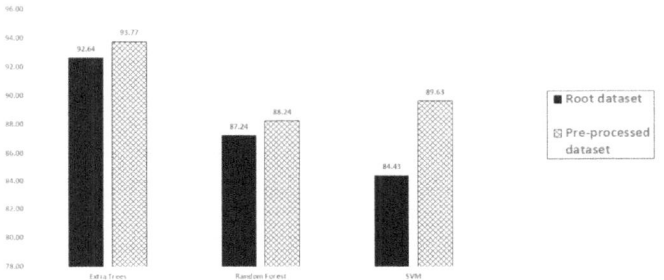

Fig. 3: The chart of values in some accuracies with different kind of input data and machine learning models

and ROC-AUC values than the other models. While the SVM model's estimate of its learning capacity is superior, its performance with this data is the worst. Yet, when comparing one component of One-vs.-Rest Specificity to another, we detect an anomalous characteristic. The SVM model has a better specificity than Random Forest, which is distinct from the three other measures. It indicates that the SVM model is more resistant to type I errors due to its unique processes for tackling classification issues.

5 Conclusion

Based on studies into the correlation between a person's glucose level and Raman spectroscopy reflected from various bodily areas, we suggested a method for assessing glucose utilizing a Raman spectrometer and a machine learning system with many models. Before the machine learning models anticipate the glucose level from the samples, the spectrometer generates Raman spectra from human samples. We utilized the Technospex uRaman - Ci spectrometer to build a dataset from glucose-mixed fluids with varying glucose concentrations. Before being labeled as primitive data, the dataset has been cleansed of noise. We also utilized feature-extraction techniques to enhance the performance of machine learning systems by reducing the size of input data. The preprocess data based on simple calculation to define data series bring valuable information. Thus, the preprocess data becomes shorter, but brings features of glucose levels. The input data for the machine learning model could be basic data or data extracted during the extraction procedure.

Before testing the accuracy of the trained models' classification of samples, we designed experiments in which machine learning algorithms extracted characteristics from the dataset. Experiments utilized three models: Extra Trees, Random Forest, and SVM model. The results proved the efficiency of machine learning on classification problems as well as the preprocessing procedures that meet our requirements. The accuracy ranges from 80% to 97%, while the extraction process increases the accuracy of each machine learning model in the same experimental dataset.

We will raise the dataset's complexity in the future by increasing the number of labels and the number of samples for each label. Non-invasive sampling methods for Raman spectroscopy data collection will also be thoroughly investigated, and we will continue to improve the machine learning models and preprocessing techniques used to extract characteristics from the dataset.

References

1. Raman, C.V., Krishnan, K.S.: A new type of secondary radiation. Nature **121**(3048), 501–502, (1928)
2. Schmid, T., Dariz, P.: Raman microspectroscopic imaging of binder remnants in historical mortars reveals processing conditions. Heritage **2**(2), 1662–1683 (2019)
3. Jun, X., et al.: Raman spectroscopy as a versatile tool for investigating thermochemical processing of coal, biomass, and wastes: recent advances and future perspectives. Energy Fuels **35**(4), 2870–2913 (2020)
4. Todaro, B., et al.: "Is Raman the best strategy towards the development of non-invasive continuous glucose monitoring devices for diabetes management?." Front. Chem.**10**, 994272 (2022) https://doi.org/10.3389/fchem.2022.994272
5. Yang, S.J., et al.: "Rapid identification of microplastic using portable Raman system and extra trees algorithm." Real-time Photonic Measurements, Data Management, and Processing V, Vol. 11555. SPIE, 2020
6. Alsariera, Y.A., Adeyemo, V.E., Balogun, A.O., Alazzawi, A.K.: AI meta-learners and extra-trees algorithm for the detection of phishing websites. IEEE Access **8**, 142532–142542 (2020)
7. Nakanishi, A., et al.: "Development of a Prediction Method of Cell Density in Autotrophic/Heterotrophic Microorganism Mixtures by Machine Learning Using Absorbance Spectrum Data." BioTech **11**(4), 46 (2022):
8. Sadat-Mohammadi, M., et al.: "Non-invasive physical demand assessment using wearable respiration sensor and random forest classifier." J.Build. Eng. **44**, 103279 (2021)
9. Khan, Z.Y., Niu, Z., Sandiwarno, S.: Deep learning techniques for rating prediction: a survey of the state-of-the-art. Artif. Intell. Rev. **54**, 95–135 (2021)
10. Shokrekhodaei, M., Cistola, D.P., Roberts, R.C., Quinones, S.: Non-invasive glucose monitoring using optical sensor and machine learning techniques for diabetes applications. HHS Public Access, IEEE Access **9**, 73029–73045 (2021)
11. Guevara, E., Torres-Galván, J.C., Ramírez-Elías, M.G., Luevano-Contreras, C., González, F.J.: Use of raman spectroscopy to screen diabetes mellitus with machine learning tools. Biomed. Opt. Express, 9(10): 4998–5010, 2018
12. Belgiu, M., Dragut, L.: Random forest in remote sensing: a review of applications and future directions. ISPRS J. Photogramm. Remote Sens. **114**, 24–31 (2016)
13. Theobald, N., et al. "Identification of unknown nanofabrication chemicals using raman spectroscopy and deep learning." IEEE Sens. J. (2023)

14. Pisner, D.A., Schnyer, D.M.: Support vector machine. Machine Learning, Chapter 6: 101–121, Academic Press (2020)

15. Sujay Raghavendra, N., Deka, P.C.: Support vector machine applications in the field of hydrology: a review. Appl. Soft Comput. **19**, 372–386 (2014)

16. Lin, J., et al.: Raman spectroscopy of human hemoglobin for diabetes detection. J. Innovative Opt. Health Sci. **7**(1), 1350051–1350056 (2014)

17. Berger, A.J., Itzkan, I., Feld, M.S.: Feasibility of measuring blood glucose concentration by near-infrared Raman spectroscopy. Spectrochim. Acta Part A Mol. Biomol. Spectrosc. **53**(2), 287–292 (1997)

18. Caduff, A., Hirt, E., Feldman, Y., Ali, Z., Heinemann, L.: First human experiments with a novel non-invasive, non-optical continuous glucose monitoring system. Biosens. Bioelectron. **19**(3), 209–217 (2003)

19. Ehsan, U., et al.: "Surface-enhanced Raman spectroscopy of centrifuged blood serum samples of diabetic type II patients by using 50KDa filter devices." Spectrochimica Acta Part A: Molecular and Biomolecular Spectroscopy **293**, 122457 (2023):

20. Zoppis, I., Mauri, G., Dondi, R.:"Kernel methods: Support vector machines." Encyclopedia of Bioinformatics and Computational Biology. Volume 1. Elsevier, 503–510 (2019)

21. Shokrekhodaei, M., et al.: "Non-invasive glucose monitoring using optical sensor and machine learning techniques for diabetes applications." IEEE Access **9**, 73029–73045 (2021)

Comprehensive Survey On Remote Sensing Image Processing Techniques for Image Classification

Thuy Thi Tran[1] 🆔 and Hiep Xuan Huynh[2(✉)] 🆔

[1] Faculty of Information Technology – Communication, University of CuuLong,
Vinh Long, Vietnam
[2] College of Information and Communication Technology,
Can Tho University, Can Tho, Vietnam
hxhiep@ctu.edu.vn

Abstract. Remote sensing technology is now being used with unprecedented high resolution, making important contributions in practical applications such as urban development, construction planning and disaster prediction. However, although many scholars have studied algorithms for remote sensing image processing, there have not been detailed summary articles to support new researchers in this field. In this paper, we present an overview of remote sensing images, types of remote sensing satellite images and related studies. Next, we briefly review the recent history of remote sensing image processing techniques. Then, list related studies on remote sensing image classification. Finally, based on the current status of research on remote sensing images, we propose some future research directions in order to provide survey references for new studies in the field of remote sensing.

Keywords: remote sensing · resolution · satellite sensors · classification · spatial point pattern

1 Introduction

Remote sensing is the process of gathering data about the earth's surface without being in contact with it [1]. This is process is done by sensing and recording emitted or reflected energy and then processing, analyzing and applying that information. Remote sensing process includes the illumination or energy source which passes through the atmosphere and interacts with the target; the electromagnetic energy emitted or scattered from the target is collected and recorded by the satellite sensors is transmitted in electronic form to a receiving and processing station where the data is processed into an image [2]. The processed image is interpreted visually or electronically or digitally to extract the information about the illuminated target. Remote sensing systems which measure reflected energy are called passive sensors, which can be used only to detect energy in the present of naturally occurring energy. This can take place only during the time when the sun is illuminating the earth [3].

P. Cong Vinh and N. Thanh Tung (Eds.): ICCASA 2023, LNICST 579, pp. 102–114, 2024.
https://doi.org/10.1007/978-3-031-58878-5_7

An active sensor provides its own energy source for illumination [2]. The sensors emit radiation which is directed towards the target to be investigated; these sensors obtain the information regardless of the time of day. In order to capture the earth's surface the sensors must be paced in a proper platform. Before it was ground-based and aircrafts platforms, nowadays satellite near-polar orbits platform provides a great contribution to remote sensing imagery.

The Multispectral satellite sensor provides digital raster images, that allow us to apply Digital Image Processing techniques to develop thematic maps of landuse/landcover classes which are essential in many remote sensing applications like forestry, agriculture, environmental studies, weather forecasting, ocean studies, archeological studies etc.

In this paper various advanced image processing techniques to convert raw satellite imagery into fine data obtained from different spatial, spectral and temporal resolutions from microwave to ultraviolet bands are discussed.

The paper is organized as follows; Sections 1, 2 deal with various resolutions of satellite sensor. Section 3 describes Satellite sensors for distinct applications. Sections 4, 5 is a study of Image Analysis which includes advance algorithms for preprocessing, enhancement, transformation and classification. Section 6 presents conclusion.

2   Remote Sensing Satellite Imagery

Remote sensing satellite image consists of Digital Numbers which represent image features such as color, brightness, wavelength, radiated energy frequency, or picture element in the image. The smallest element on an image is called pixel. A digital image consists of pixels which are arranged in rows and columns commonly known as a raster image. The information content and dimensions of these pixels depend on the resolution of the image. Figure 1 shows various sensor resolutions.

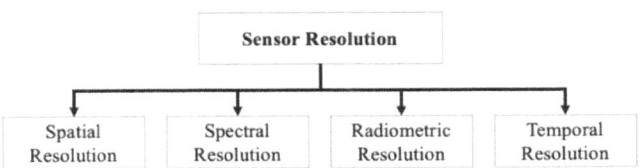

Fig. 1. Satellite Sensor Resolutions

2.1 Spatial Resolution

The detail of an image depends on spatial resolution of the sensor. If the spatial resolution is 10 m, it means that each pixel denotes an area of 10 m/10 m on the ground surface. Higher the resolution of an image, finer details is more clearly-visible and cover less ground area. Lower the resolution of an image, details are not clearly visible but it covers larger total ground area.

Yu Huang et al., [4] indicated that as the resolution decreases, the accuracy of land-slide detection also decreases. The overall landslide area detection rate of UAV imagery can reach 82.17%, while that of GF-6 and Landsat 8 imagery is only 52.26% and 48.71%. The landslide quantity detection rate of UAV imagery can reach 99.07%, while that of GF-6 and Landsat 8 images is only 48.71% and 61.05%. In addition, for each landslide detected, little difference is found in large-scale landslides, and it becomes more difficult to correctly detect small-scale landslides as the resolution decreases. For example, landslides under 100 m2 could not be detected from a Landsat 8 satellite image.

2.2 Spectral Resolution

Spectral resolution of a sensor is the ability to define fine wavelength intervals in an Electromagnetic spectrum. The details of an image also depend on responses of Electromagnetic Radiation incident on an object over distinct wavelength ranges, for example, the classification of vegetation and water is usually be separated in a broad wavelength range i.e., visible and near infrared wavelengths, to distinguish different rocks needs finer wavelength range within the band to separate them. So higher the spectral resolution, narrower the wavelengths range of a particular band.

Zou et al. [5] CCC-sensitive and MTA-insensitive satellite broadband vegetation indices are developed for crop canopy chlorophyll content estimation. The most efficient broadband vegetation indices for four satellite sensors (Sentinel-2, RapidEye, WorldView-2 and GaoFen-6) with red edge channels were identified (in the context of various vegetation index types) using simulated satellite broadband reflectance based on field measurements and validated with PROSAIL model simulations. The results indicate that developed vegetation indices present strong correlations with CCC and weak correlations with MTA, with overall R2 of 0.76–0.80 and 0.84–0.95 for CCC and R2 of 0.00 and 0.00–0.04 in the field measured data and model simulations, respectively. The best vegetation indices identified in this study are the soil-adjusted index type index SAI (B6, B7) for Sentinel-2, Verrelts's three-band spectral index type index BSI-V (NIR1, Red, Red Edge) for WorldView-2, Tian's three-band spectral index type index BSI-T (Red Edge, Green, NIR) for RapidEye and difference index type index DI (B6, B4) for GaoFen-6. The identified indices can potentially be used for crop CCC estimation across species and seasonality.

Zhang et al. [6] demonstrated a novel oceanic triple-field-of-view (FOV) high-spectral-resolution lidar (HSRL) with an iterative retrieval approach. This technique provides, for the first time, comprehensive, continuous, and vertical measurements of seawater absorption coefficient, scattering coefficient, and slope of particle size distribution, which are validated by simulations and field experiments. Furthermore, it depicts valuable application potentials in the accuracy improvement of seawater classification and the continuous estimation of depth-resolved particulate organic carbon export. The triple-FOV HSRL with high performance could greatly increase the knowledge of seawater constituents and promote the understanding of marine ecosystems and biogeochemistry.

2.3   Radiometric Resolution

Radiometric resolution enables us to recognize high and low level contrast objects in an image. Radiometric resolution describes the information about image brightness, contrast, illumination variations and other details of an image.

Verde et al. [7] concluded that high radiometric resolution data does not always lead to higher and more accurate classification results, although in certain cases the classification results of higher radiometric resolution data are more accurate than lower radiometric resolution data. The study ran experiments on remote sensing datasets with radiometric resolution at three different locations. The bagging classification method used gives low radiometric resolution. The study also confirmed that image content retrieval and processing time also depend on the radiometric resolution of the image. Experiments also showed no correlation between classification based on radiometric resolution and classification based on texture bands and spectral indices.

2.4   Temporal Resolution

The time taken by a satellite to revisit the same area with same viewing angle is referred as absolute temporal resolution. It refers to the length of time it takes for a satellite to complete one entire orbit cycle. Temporal data plays a very important role in remote sensing applications like monitoring vegetation changes, flood occurrence, deforestation, urban development etc., Spectral resolution varies with time and is identified by multi temporal imagery. Temporal resolution depends on many factors like satellite sensor capability, latitude and swath overlap.

Wang et al. [8] integrated multisource remote sensing, including satellite altimetry and optical and synthetic aperture radar (SAR) images, to generate weekly water levels and water storages of nine largest reservoirs on the main stem of the LMR from 2017 to 2021. Specifically, partial surface water extent (SWE) of reservoirs was extracted from Sentinel-1 SAR images and digital elevation models (DEMs), using Random Forest algorithms trained by partial SWE derived from Sentinel-2 optical images, showing an overall accuracy higher than 95%. Based on the partial SWE and water level estimates from ICESat-2 and Global Ecosystem Dynamics Investigation (GEDI, International Space Station-based) data, the relationships between water levels and partial SWE were derived to convert partial SWE into water level time series. Furthermore, water storage time series of the nine reservoirs were obtained from water level time series and hypsometric functions derived from SRTM DEMs that were corrected by ICESat-2 data to remove systematic errors. For the Xiaowan Reservoir on the Lancang River, there is close agreement between remote sensing-derived water levels and in-situ water levels in terms of a normalized RMSE lower than 5%. Results indicate that multisource remote sensing has large potential for high-temporal-resolution monitoring of reservoir water levels and water storage.

3   Satellite Sensors

Satellites provide remote sensing imagery which are commonly used today. The unique characteristics of Satellites make them particularly useful for remote sensing of the Earth's surface.

3.1 Thermal Sensors

To measure the surface temperature and thermal properties of a target object on the ground surface, thermal sensors are used that detects the reflected radiation from the target object.

Further contributions to research in the field of land surface temperature (LST) calculation using low-altitude thermal infrared (TIR) remote sensing image data. Yafei Wu et al. [9] evaluated the small-scale urban thermal environment by proposing a block-scale land surface temperature retrieval model with high spatial and thermal resolution. Experimental data using multi-source remote sensing images. The results were compared with the land surface temperature measured in the field area with the proposed model's accuracy of 0.09K. The results of the study can be used to quantitatively analyze land surface temperature patterns in urban areas at the block scale.

3.2 Airborne and Space-Borne Sensors

Airborne remote sensing are one time operations. Here, sensors are mounted on aircrafts that provides images with high spatial resolution but it covers less ground area. Space-borne remote sensing provides continuous monitoring of earth's surface. Here, the sensors are placed on space shuttles or satellites. It covers larger earth's surface with less spatial resolution.

Hirschmugl et al. [10], evaluated forest structure based on spaceborne and airborne Lidar data in a natural forest in an Austrian Alp. Data to run waveform experiments. Number of layers and foliage height height diversity are two indicators that have been used to explain the vertical structure of natural forests. In general, the overall accuracy of the research results is affected by natural forests with high vertical structures and rugged terrain. However, this is also a study that can provide future research directions on assessing forest structure using waveform data. Future studies can run experiments on managed forest areas with a simpler structure to reevaluate the parameters used in forest structure analysis, foliage height and number of layers.

This study [11] conducted a quantitative retrieval and validation of the Leaf Area Index (LAI) profiles using terrestrial and airborne laser scanning (TLS and ALS) and spaceborne GEDI data over a deciduous needleleaf forest site in northern China. The vertical LAI profile was estimated in the field using an upward digital hemispherical photography (DHP) attached to a portable measurement system in 2020 and 2021. A suite of new LiDAR indices combining both LiDAR return number and return intensity was explored for the LAI profile estimation. All LAI profiles obtained from the DHP, TLS, ALS, and GEDI during the leaf-on season and leaf-off season were compared. The DHP shows a good agreement with the TLS LAI profiles ($R2 = 0.97$). The LAI profile derived from the ALS data using the combined light penetration index (LPIRI) agrees well ($R2 \geq 0.86$) with the DHP, TLS, and GEDI estimates. In general, the LPIRI is advantageous for regional LAI profile mapping from ALS. The GEDI cumulative LAI corresponds well with the DHP during the leaf-on season ($R2 = 0.90$, RMSE $= 0.23$), but underestimates during the leaf-off season ($R2 = 0.70$, RMSE $= 0.14$, bias $= -0.13$). The underestimation is attributed to the higher canopy and ground reflectance ratio ($\rho v/\rho g$) assigned in the algorithm and the height discrepancy between the GEDI and field measurements. For the GEDI LAI profile product, further validation and improvement are necessary for other biome types and landscape conditions, especially during the leaf-off season.

4   Image Analysis

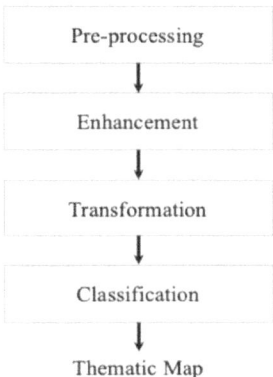

Fig. 2. Remote Sensing Analysis

In order to make good use of remote sensing data, we must be able to extract meaningful information from the image by applying proper processing techniques. Remote sensing images can also be represented in a computer as arrays of pixels, with each pixel corresponding to a digital number [DN], represents the brightness level of that pixel in an image. Figure 2 is a general architecture of remote sensing image analysis.

Image analysis systems can be categorized into following: (i) Pre-processing, (ii) Image Enhancement, (iii) Image Transformation, (iv) Image Classification and Analysis.

4.1   Pre-processing

Pre-processing functions involves the operations required prior to the main data analysis and consists of processes aimed at geometric correction, radiometric correction and atmospheric corrections to improve the ability to interpret the image components qualitatively and quantitatively. These process correct the data for sensor irregularities and removing (radiometric corrections) unwanted sensor distortion or atmospheric noise.

According to the imaging characteristics of Night time light (NTL) images, this paper [12] proposes to take high-precision road network data as geometric reference, extract control points by automatic matching between NTL images and road network data, and then realize geometric correction of NTL images. Taking the Luojia 1 -01 (LJ 1–01) satellite NTL image as an example, the experimental verification shows that the accuracy of geometric correction based on road network control can reach the sub-pixel level, which verifies the feasibility of the proposed method. This paper verifies the rationality of using the road network as the benchmark data for NTL images, provides a feasible idea for subsequent scholars to study the geometric processing of NTL images, and ensures the geometric quality of data for the application of multi-temporal NTL images.

Homomorphic filtering is employed to enhance thermal infrared (TIR) image details and the modified RIFT algorithm is proposed to achieve TIR-visible image registration [13]. Different from using MIM for feature description in RIFT, the proposed modified RIFT uses the novel binary pattern string to descriptor construction. With sufficient and uniformly distributed ground control points, the two-step orthorectification framework, from SDGSAT-1 TIS L1A image to L4 orthoimage, are proposed in this study. The first experiment, with six TIR-visible image pairs, captured in different landforms, is performed to verify the registration performance, and the result indicates that the homomorphic filtering and modified RIFT greatly increase the number of corresponding points. The second experiment, with one scene of an SDGSAT-1 TIS image, is executed to test the proposed orthorectification framework. Subsequently, 52 GCPs are selected manually to evaluate the orthorectification accuracy. The result indicates that the proposed orthorectification framework is helpful to improve the geometric accuracy and guarantee for the subsequent thermal infrared applications.

Radiometric correction is one of the most important preprocessing parts of unmanned aerial vehicle (UAV) multispectral remote sensing data analysis and application [14]. In this article, a back propagation (BP) neural network-based radiometric correction method (BPNNRCM) considering optimal parameters was proposed. In the aspect of accuracy and robustness, the absolute errors of test and cross-validation images' surface reflectance obtained by the BPNNRCM were all less than 0.054. The BPNNRCM had smaller average absolute error (0.0141), mean squared error (0.0003), mean absolute error (0.0141) and mean relative error (7.1%) comparing with empirical line method and radiative transfer model.

Pahlevan et al. [15] summarize their results using performance matrices guiding the satellite user community through the Optical Water Types specific relative performance of Atmospheric correction processors. Their analysis stresses the need for better representation of aerosols, particularly absorbing ones, and improvements in corrections for sky- (or sun-) glint and adjacency effects, in order to achieve higher quality downstream products in freshwater and coastal ecosystems.

Quinten et al. [16], proposed a dark spectrum fitting (DSF) atmospheric correction model based on high-resolution optical remote sensing images such as Sentinel-2 and Landsat, for underwater applications. The model includes two processing processes: image brightness correction and automatic layering of panoramic images. The proposed solution used multiple dark targets in the secondary scene to construct a dark spectral range used to estimate atmospheric path reflectance based on the best-fit aerosol model. This model is completely automatic so it can be used to process the entire study area, specifically the North Sea. DSF will select the most appropriate range to overcome the problem of amplifying lighting effects in atmospheric correction. From the SWIR bands, the reflectance of sunlight can be estimated for calibration and use in low areas such as deep sea.

4.2 Enhancement

To make image easier for visual interpretation Enhancements are used. The advantage of digital imagery is that it allows us to manipulate the digital pixel values in an image. Although radiometric corrections for illumination, atmospheric influences, and sensor characteristics may be done prior to distribution of data to the user, the image may still not be optimized for visual interpretation. Image Enhancement methods are of four types: (i) Radiometric Enhancement; (ii) Spatial Enhancement; (iii) Spectral Enhancement; (iv) Geometric Enhancement.

One of the major disadvantages of waveform airborne laser scanners is the loss of signal strength of the data due to the path of the laser pulse through objects such as tree canopies or buildings. Richter et al., [17] proposed a model for radiometric enhancement of full-waveform airborne laser scanner data. This research can be used in environmental applications to represent the volume of objects that need to be studied. Another important contribution of the study is the ability to provide information about the structure of plants using radiometric enhancement techniques of airborne laser scanners.

Yang et al., [18] proposed a new model based on spatially enhanced UNet to handle the global road network by combining dense connection points and spatial convolutional neural network. The proposed solution is the integration of a structural conservation model and previously collected road surface information to be able to predict successive events in space. Experimental results have shown that the proposed solution has better performance than previous road solutions.

This article [19] reviews and discusses the most important algorithms relevant to this area of research between 2002 and 2022, along with the most frequently used datasets, HSI sensors, and quality metrics. Metaanalysis are drawn based on the aforementioned information, which is used as a foundation that summarizes the state of the field in a way that bridges the past and the present, identifies the current gap in it, and recommends possible future directions.

The geometric enhancement of the OpenStreetMap (OSM) road network using a standard national map as a reference. Belhouari et al., [20] use two transformation methods, the global transformation and the local transformation. The application of this approach in the geometric enhancement / correction where each node of the OSM network will have a newly calculated position. Both approaches have been tested in the region of Oran in Algeria as testing example.

4.3 Transformation

Image transformations typically involve the manipulation of multiple bands of data, whether from a single multispectral image or from two or more images of the same area acquired at different times (i.e. multitemporal image data). Either way, image transformations generate new images from two or more sources that highlight particular features or properties of interest, better than the original input images. Image transformation includes basic arithmetic operations like band Arithmetic operations are performed on two or more co- registered images of same geographical area. They may have different spectral band from a single multispectral data or it may have individual band of different time series data set.

The spectral transformation highlighted the characteristics of spectral curves and improved the relationship between spectral reflectance and anthocyanin, and the remote sensing model based on the first-order differential spectrum portrayed the best estimation accuracy (R2c = 0.91; R2v = 0.51) [21].

Zhang et al., [22] improved the remote sensing-based soil salinity content extraction from the Landsat 8, Digital elevation model and HJ-1A CCD satellite data using the Cuckoo Search Algorithms-Support Vector Machines model. The analysis of soil and vegetation factors shows that the first three principal components cumulative variance contributed 99.69% on the raw remote sensing image, while the first two principal components cumulative variance contributed 88.01% and 85.28% on the first- and second-order differential transformation remote sensing images, respectively.

5 Image Classification and Analysis

Image classification is an important part of the remote sensing, image analysis and pattern recognition. Based on the idea that different feature types on the earth's surface have a different spectral reflectance and remittance properties, their recognition is carried out through the classification process.

This study [23] presented the performance of the Mangrove Vegetation Index (MVI) and image classification algorithms, embedded in the Google Earth Engine, applied to Landsat-8 and Sentinel-2 data, to map tracts of mangroves in Aracaju (Sergipe, Brazil). Results reveal that the Cobweb clustering algorithm applied to MVIderived from Landsat-8 data favors reliable and practical mangrove mapping, considering the broad diversity of vegetation conditions in this habitat.

To improve the global cloud detection performance for Landsat satellite imagery, Pang et al., [24] used a combination of convolutional neural network models. Landsat satellite imagery provides discrete spectral channels from short wavelengths (green, blue, red) to visible infrared wavelengths through its surface radiometric sensors of objects on Earth. The experimental results were re-evaluated with the Landsat 8 Bio dataset and it was determined that the cloud of the extended UNet model had the best results among the estimated models.

This review [25] provided an overview of Earth Observation data, machine learning and state-of-the-art deep learning techniques that are currently being used to quantify above-ground carbon, below-ground carbon, and soil carbon stocks of mangroves, seagrasses and saltmarshes ecosystems. Some key limitations and future directions for the potential use of data fusion combined with advanced machine learning, deep learning, and metaheuristic optimisation techniques for quantifying blue carbon stocks are also highlighted. In summary, the quantification of blue carbon using remote sensing and machine learning approaches holds great potential in contributing to global efforts towards mitigating climate change and protecting coastal ecosystems.

The urbanization process greatly affects the global warming process. The identification of urban areas plays an increasingly important role. One of the solutions that is continuous in space and time and has a low cost to support this problem is to use remote sensing images. Chen et al., [26] used Landsat 8 satellite images to extract information about land surface temperature, combined with OpenStreetMap to locate urban areas, using point of interest to determine the data area that needs attention for processing. A deep learning algorithm is a random forest used to re-evaluate urban areas with respect to the temperature environment of the land surface. The assessment results are correct for areas along the Hunhe River with gradually decreasing land surface temperatures, and higher temperatures in urban central areas. From the results of this study, future researchers in the field of processing land surface temperature from remote sensing images can do further research to provide results to support functional sectors in the process of urbanization in a reasonable way to combat climate change.

Deep Convolutional Embedded Clustering (DCEC) is a new unsupervised deep learning method, which was used by Maarten et al., [27] to generate a landscape typology for Switzerland. This method encodes the input image into a hidden layer. This hidden layer is used to classify images into separate clusters such as demographic classes, terrain, flora and fauna ecology, etc. The results of the study were successfully run experiments to distinguish 45 types of continuous landscapes with input data from remote sensing images. This is a promising solution for future researchers in the field of land systems, geology and landscape classification.

The focus of this study [28] was to investigate the application of hyperspectral remote sensing and deep learning (DL) for real-time ore and waste classification. Hyperspectral images of several meters of drill core samples from a silver ore deposit labeled by a site geologist as ore and waste material were used to train and test the models. A DL model was trained on the labels generated by a spectral angle mapper (SAM) machine learning technique. The performance on ore/waste discrimination of three classifiers (supervised DL and SAM, and unsupervised k-means clustering) was evaluated using Rand Error and Pixel Error as disagreement analysis and accuracy assessment indices. The results showed that the DL method outperformed the other two techniques. The performance of the DL model reached 0.89, 0.95, 0.89, and 0.91, respectively, on overall accuracy, precision, recall, and F1 score, which indicate the strong capability of the DL model in ore and waste discrimination. An integrated hyperspectral imaging and DL technique has strong potential to be used for practical and efficient discrimination of ore and waste in a near real-time manner.

Xu et al., [29] proposed an unsupervised domain adaptation model to solve the difficulty caused by the large distribution difference between the source and target domains. The purpose of the proposed model is that during the conversion process, it is necessary to keep the data in the source domain intact for classification in the target domain, especially for unlabeled data.

Classification methods based on thresholds of vegetation indices do not accurately estimate the flooded area in areas with heterogeneous water surfaces. Foroughnia et al., [30] used synthetic aperture radar and multispectral data and machine learning methods for unsupervised and supervised classification to classify flooded areas. This solution overcame problems with water clarity and emerging vegetation.

Stromann et al., [31] used the computational power of Google Earth Engine and Google Cloud Platform to generate an oversized feature set in which we explore feature importance and analyze the influence of dimensionality reduction methods to object-based land cover classification with Support Vector Machines. They propose a methodology to extract the most relevant features and optimize an SVM classifier hyperparameters to achieve higher classification accuracy. The proposed approach is evaluated in two different urban study areas of Stockholm and Beijing. Despite different training set sizes in the two study sites, the averaged feature importance ranking showed similar results for the top-ranking features. In particular, Sentinel-2 NDVI, NDWI, and Sentinel-1 VV temporal means are the highest ranked features and the experiment results strongly indicated that the fusion of these features improved the separability between urban land cover and land use classes. Overall classification accuracies of 94% and 93% were achieved in Stockholm and Beijing study sites, respectively. The test demonstrated the viability of the methodology in a cloud-computing environment to incorporate dimensionality reduction as a key step in the land cover classification process, which they consider essential for the exploitation of the growing Earth observation big data.

6 Conclusion

Remote sensing is going mainstream, both in the business and personal worlds, and has the potential to lead us into the metaverse. Within the framework of this article, 31 articles have been selected and analyzed, including different cases of resolution, sensing of remote sensing images, analysis and classification of remote sensing images. Particular emphasis is placed on the various machine learning methods (supervised and unsupervised) adopted by the research community in conjunction with Earth Observation data, used as well as problem statements. The general comment of this study is that there is no standardized approach to general application for remote sensing image processing to bring the most optimal efficiency. We propose spatial point pattern approaches to further improve the accuracy, efficiency, and applicability of remote sensing image classification in the combination of computer science and business. Specifically, we use the spatial point pattern technique to classify the locations of restaurants, hotels or residential areas so that investors can strategically deploy a retail agency system with appropriate locations to improve business efficiency as much as possible.

References

1. Gibson, P.J., Power, C.H., Keating, J.: Introductory Remote Sensing: Principles and Concepts. Routledge (2013). https://doi.org/10.4324/9780203714522
2. Richards, J A.: Remote Sensing Digital Image Analysis, Sixth Edition, Springer (2022).
3. Revanna, S., Deepa, P., Venugopal, K.R.: Remote sensing satellite image processing techniques for image classification. Int. J. Comput. Appl. **161**, 24–37 (2017). https://doi.org/10.5120/ijca2017913306
4. Huang, Y., et al.: How spatial resolution of remote sensing image affects earthquake triggered landslide detection: an example from 2022 luding earthquake, Sichuan, China. Land **12**, 681 (2023). https://doi.org/10.3390/land12030681
5. Zou, X., Jin, J., Mõttus, M.: Potential of satellite spectral resolution vegetation indices for estimation of canopy chlorophyll content of field crops: mitigating effects of leaf angle distribution. Remote Sens. **15**, 1234 (2023). https://doi.org/10.3390/rs15051234

6. Zhang, K., et al.: Comprehensive, continuous, and vertical measurements of seawater constituents with triple-field-of view high-spectral-resolution lidar. Research **6**, 0201 (2023) https://doi.org/10.34133/research.0201

7. Verde, N., Mallinis, G., Tsakiri-Strati, M., Georgiadis, C., Patias, P.: Assessment of radiometric resolution impact on remote sensing data classification accuracy. Remote Sens. **10**, 1267 (2018). https://doi.org/10.3390/rs10081267

8. Wang, Y., Long, D., Li, X.: High-temporal-resolution monitoring of reservoir water storage of the Lancang-Mekong River. Remote Sens. Environ. **292**, 113575 (2023). https://doi.org/10.1016/j.rse.2023.113575

9. Wu, Y., Shan, Y., Lai, Y., Zhou, S.: Method of calculating land surface temperatures based on the low-altitude UAV thermal infrared remote sensing data and the near-ground meteorological data. Sustain. Cities Soc. **78**, 103615 (2022). https://doi.org/10.1016/j.scs.2021.103615

10. Hirschmugl, M., Lippl, F., Sobe, C.: Assessing the vertical structure of forests using airborne and spaceborne LiDAR data in the Austrian alps. Remote Sens. **15**, 664 (2023). https://doi.org/10.3390/rs15030664

11. Wang, Y., Fang, H., Zhang, Y., Li, S., Pang, Y., Ma, T.: Retrieval and validation of vertical LAI profile derived from airborne and spaceborne LiDAR data at a deciduous needleleaf forest site. GISci. Remote Sens. **60**(1) (2023) https://doi.org/10.1080/15481603.2023.2214987

12. Zhang, L., Zhang, G., Liu, W., Li, Z., Xie, T.: Geometric correction of Luojia 1-01 nighttime light image based on road network. J. Imaging Sci. Technol. **67**(2), 1–12 (2023). https://doi.org/10.2352/J.ImagingSci.Technol.2023.67.2.020401

13. Chen, J., et al.: A TIR-visible automatic registration and geometric correction method for SDGSAT-1 thermal infrared image based on modified RIFT. Remote Sens. **14**, 1393 (2022). https://doi.org/10.3390/rs14061393

14. Zhang, Y., et al.: A back propagation neural network-based radiometric correction method (BPNNRCM) for UAV multispectral image. IEEE J. Sel. Top. Appl. Earth Observations Remote Sens. **16**, 112–125 (2023). https://doi.org/10.1109/JSTARS.2022.3223781

15. Pahlevan, N., et al.: ACIX-Aqua: A global assessment of atmospheric correction methods for Landsat-8 and Sentinel-2 over lakes, rivers, and coastal waters. Remote Sens. Environ. **258**, 112366 (2021). https://doi.org/10.1016/j.rse.2021.112366

16. Vanhellemont, Q.: Adaptation of the dark spectrum fitting atmospheric correction for aquatic applications of the landsat and sentinel-2 archives. Remote Sens. Environ. **225**, 175–192 (2019). https://doi.org/10.1016/j.rse.2019.03.010

17. Richter, K., Maas, H.-G.: Radiometric enhancement of full-waveform airborne laser scanner data for volumetric representation in environmental applications. ISPRS J. Photogramm. Remote Sens. **183**, 510–524 (2022). https://doi.org/10.1016/j.isprsjprs.2021.10.021

18. Yang, M., Yuan, Y., Liu, G.: SDUNet: road extraction via spatial enhanced and densely connected UNet. Pattern Recogn. **126**, 108549 (2022). https://doi.org/10.1016/j.patcog.2022.108549

19. Aburaed, N., Alkhatib, M.Q., Marshall, S., Zabalza, J., Ahmad, H.A.: A review of spatial enhancement of hyperspectral remote sensing imaging techniques. IEEE J. Sel. Top. Appl. Earth Observations Remote Sens. **16**, 2275–2300 (2023). https://doi.org/10.1109/JSTARS.2023.3242048

20. Belhouari, F.Z., Boukerch, I., Siyoucef, K.: Geometric Enhancement of the openstreetmap road network. ISPRS Ann. Photogrammetry Remote Sens. Spat. Inf. Sci. **V-4–2021**, 33–39 (2021). https://doi.org/10.5194/isprs-annals-V-4-2021-33-2021

21. Luo, L., Chang, Q., Gao, Y., Jiang, D., Li, F.: Combining different transformations of ground hyperspectral data with unmanned aerial vehicle (UAV) images for anthocyanin estimation in tree peony leaves. Remote Sens. **14**, 2271 (2022). https://doi.org/10.3390/rs14092271

22. Zhang, F., et al.: Retrieval of soil salinity based on multi-source remote sensing data and differential transformation technology. Int. J. Remote Sens. **44**(4), 1348–1368 (2023). https://doi.org/10.1080/01431161.2023.2179900

23. Rodrigues, F., de Souza Filho, C.R., Del Papa, R., Scafutto, M., Lassalle, G.: Mangrove mapping strategies using google earth engine and landsat8 and sentinel-2 imagery data, Anais do Simposio Brasileiro de Sensoriamento Remoto (2023)

24. Pang, S., Sun, L., Tian, Y., Ma, Y., Wei, J.: Convolutional neural network-driven improvements in global cloud detection for landsat 8 and transfer learning on sentinel-2 imagery. Remote Sens. **15**, 1706 (2023). https://doi.org/10.3390/rs15061706

25. Pham, T.D., et al.: Advances in Earth observation and machine learning for quantifying blue carbon. Earth Sci. Rev. **243**, 104501 (2023). https://doi.org/10.1016/j.earscirev.2023.104501

26. Chen, Y., Yang, J., Yang, R., Xiao, X., Xia, J.C.: Contribution of urban functional zones to the spatial distribution of urban thermal environment. Build. Environ. **216**, 109000 (2022). https://doi.org/10.1016/j.buildenv.2022.109000

27. van Strien, M.J., Grêt-Regamey, A.: Unsupervised deep learning of landscape typologies from remote sensing images and other continuous spatial data. Environ. Model. Softw. **155** (2022) https://doi.org/10.1016/j.envsoft.2022.105462

28. Abdolmaleki, M., Consens, M., Esmaeili, K.: Ore-waste discrimination using supervised and unsupervised classification of hyperspectral images. Remote Sens. **14**, 6386 (2022). https://doi.org/10.3390/rs14246386

29. Xu, M., Wu, M., Chen, K., Zhang, C., Guo, J.: The eyes of the gods: a survey of unsupervised domain adaptation methods based on remote sensing data. Remote Sens. **14**, 4380 (2022). https://doi.org/10.3390/rs14174380

30. Foroughnia, F., Alfieri, S.M., Menenti, M., Lindenbergh, R.: Evaluation of SAR and optical data for flood delineation using supervised and unsupervised classification. Remote Sens. **14**, 3718 (2022). https://doi.org/10.3390/rs14153718

31. Stromann, O., Nascetti, A., Yousif, O., Ban, Y.: Dimensionality reduction and feature selection for object-based land cover classification based on sentinel-1 and sentinel-2 time series using google earth engine. Remote Sens. **12**, 76 (2020). https://doi.org/10.3390/rs12010076

Item-Based Energy Clustering Recommendation

Tu Cam Thi Tran[1] (ID), Lan Phuong Phan[2], and Hiep Xuan Huynh[2(✉)]

[1] Vinh Long University of Technology Education, Vinh Long province, Vietnam
tuttc@vlute.edu.vn
[2] Can Tho University, Can Tho City, Vietnam
{pplan,hxhiep}@ctu.edu.vn

Abstract. Previous recommendation systems have focused on algorithms to make the recommendations based on the individual items. However, in many areas, the introduction about a cluster of the items based on the general characteristics of the item is more important than just focusing on the individual items. In this paper, we have proposed a new approach for the recommendation system, the proposed method uses the energy distance to group the items with similar properties or characteristics into a cluster, then based on the item clusters to give the most suitable recommendations for the users. In addition, the methods based on error (MAE_(c)) and accuracy (Precison_(c)-Recall_(c)) are also selected to evaluate the reliability of the new proposed model on two popular datasets Jester5k and MovieLens100k. Besides, the proposed model is also compared with two item-based collaborative filtering models using the Cosine and Pearson measures in "rrecsys" package and three item-based collaborative filtering models using the Matching, Euclidean and Karypis measures in "recommenderlab" package. The experimental results have shown that the proposed model is better than the compared models.

Keywords: Item-based · Energy distance · Clustering recommendation · Recommendation system · Item clusters

1 Introduction

A recommendation system [1] is a decision support system that provides recommendations by predicting user preferences. In addition, a group recommendation system [2, 3, 5] analyzes the interests of the group members and makes a final decision, which will be accepted by all members. The Energy [4, 16] measures the distance between the distributions of random vectors. The dimensions of those vectors are not certainly equal. Energy distance is also widely applied in research [16] such as: testing independence by distance covariance, goodness-of-fit, generalizations of clustering algorithms, change point analysis, feature selection, etc.

Previously studied energy-based recommendation systems [3, 4] were mainly focused on giving the recommendations based on individual items or the user group-based recommendation system [3]. In many areas, introduction about a cluster of the items [12] (e.g. movie cluster, home appliances cluster, etc. These clusters are based on

P. Cong Vinh and N. Thanh Tung (Eds.): ICCASA 2023, LNICST 579, pp. 115–125, 2024.
https://doi.org/10.1007/978-3-031-58878-5_8

general characteristics of the item to group), this is more important than just focusing on the individual items [11]. However, the relationship between the item clusters using energy distance in the recommendation system has not yet been considered in all these studies.

In this article, we propose a new recommendation model that considers relationships among the item clusters using energy distance. This approach is made on the basis of determining the energy relationship between the item cluster in pairs. In addition, we used an accuracy-based evaluation method (Precision_(c), Recall_(c)) and using error-based evaluation methods (MAE_(c) to evaluate the proposed model, and compared it with two item-based collaborative filtering models using the Cosine and Pearson measures in "rrecsys" package and three item-based collaborative filtering models using the Matching, Euclidean and Karypis measures in "recommenderlab" package.

The structure of the article is organized as follows. Section 2 presents related work, presenting: the clustering recommendation based on the items, the energy approach. Sect 3 shows the methods to be used for evaluating the clustering recommendation models (the accuracy-based evaluation method, the error-based evaluation method). Sect 4 depicts the proposed model that uses the relationship among the item clusters and the energy distance. Sect 5 shows the experiment results on the "recommenderlab" and "rrecsys" package for both the Jester5k datasets and MovieLens100k datasets. Section 6 is the conclusion.

2 Related Work

2.1 The Clustering Recommendation Based on the Items

The clustering recommendation systems [3, 8, 9] are clustered based on the types of cluster to which the system recommends. Clusters can be mainly based on the interactions among the members of the cluster.

Table 1. Ratings matrix for the item cluster.

Cluster	Cluster_members	Rating $r(i_j, u_k)$			
		u_1	u_2	...	u_k
$c1$	i_1	?	$r(i_j, u_k)$	...	?
	i_2	$r(i_j, u_k)$	?	...	$r(i_j, u_k)$
	i_3	$r(i_j, u_k)$	?	...	?
$c2$	i_4	$r(i_j, u_k)$	?	...	?
	i_5	$r(i_j, u_k)$	$r(i_j, u_k)$	...	$r(i_j, u_k)$
	i_6	$r(i_j, u_k)$	$r(i_j, u_k)$	...	$r(i_j, u_k)$
$c3$	i_7	?	?	...	?
	i_8	$r(i_j, u_k)$	?	...	?
	i_9	$r(i_j, u_k)$	?	...	?
cn	...	...	...	...	...
	i_j	$r(i_j, u_k)$	?	...	$r(i_j, u_k)$

The clustering is a process used to partition the data into a cluster. The data with the same characteristics, or the similar properties, they will be grouped into a cluster. The number of the clusters will be less than the number of individual items of the original data. The characteristic of the clustering method is to reduce the amount of data to be compared, to save time for the recommendation. For example, **Table 1** presented the ratings matrix for the item cluster.

2.2 The Energy Approach

Energy distance.
Energy distance [7, 16] is a statistical distance between the observed variables. The concept is based on the notion of Newton's gravitational potential energy, which is a function of the distance between two bodies in a gravitational space. Energy distance is applied to random vectors, where these random vectors have an unlimited size. Let $I_1 = I_{11}, I_{12}, \ldots, I_{1n}$ and $I_2 = I_{21}, I_{22}, \ldots, I_{2m}$ be independent random vectors in Euclidean space. The energy distance between I_1 and I_2 is define as:

$$\varepsilon_{n,m}(I_1, I_2) = 2E|I_1 - I_2|_d - E|I_1 - I_1'|_d - E|I_2 - I_2'|_d \tag{1}$$

In (1), a random variable $I_{1'}$ (or $I_{2'}$) represents a copy, which is independent and distributed like I_1 (or I_2).

The potential energy (shortly, energy) of the independent random variables I_1 and I_2 is defined by distance function ε as the follow:

Where:

$$E|I_1 - I_2|_d = \frac{1}{nm} \sum_{i=1}^{n} \sum_{j=1}^{m} |I_{1i} - I_{2j}| \tag{2}$$

$$E|I_1 - I_1'|_d = \frac{1}{n^2} \sum_{i=1}^{n} \sum_{j=1}^{n} |I_{1i} - I_{1j}| \tag{3}$$

$$E|I_2 - I_2'|_d = \frac{1}{m^2} \sum_{i=1}^{m} \sum_{j=1}^{m} |I_{2i} - I_{2j}| \tag{4}$$

Advantages of energy distance include: Energy distance is very easy to compute, it is consistent and require no distributional assumptions other than finite first moments.

The energy between the clusters.
The energy [6, 16] between the clusters was calculated from the data in the rating matrix (U x I x R), where each row is a multivariate observation.

Energy distance is used to measure the statistical distance between two clusters, and to search for the best partition between clusters.

The energy distance between two clusters C_i, C_j of size n_i, m_j is the energy distance $e(C_i, C_j)$, defined by:

$$e(C_i, C_j) = \frac{n_i m_j}{n_i + m_j} [2G_{ij} - G_{ii} - G_{jj}] \tag{5}$$

where

$$G_{ij} = \frac{1}{n_i m_j} \sum_{p=1}^{n_i} \sum_{q=1}^{m_j} \|I_{ip} - I_{jq}\|^{\alpha} \tag{6}$$

In (5), C_i is the cluster i, C_j is the cluster j; I_{ip} denotes the p-th observation in the i-th cluster, The exponent alpha should be in the interval (0,2], $\|.\|$ denotes Euclidean norm.

2.3 Evaluation Approach for the Clustering Recommendation

To evaluate the clustering recommendation system [13, 14], the accuracy-based evaluation method (the precision_(c), recall_(c)), and the error-based evaluation method (MAE_(c)), both are used.

The accuracy-based evaluation method.
Precision [13] is the fraction of the number of relevant recommended items (true positives) in relation to the total number of recommended items.

$$precision\_(c) = \frac{|predicted_k(c) \cap relevant(c)|}{k} \tag{7}$$

Recall [13] is the fraction of the number of relevant recommended items in relation to the number of all relevant items.

$$recall\_(c) = \frac{|predicted_k(c) \cap relevant(c)|}{relevant(c)} \tag{8}$$

where, k is the length of the list of recommended items and c is cluster.
  $predicted_k(c)$ shows a list of k items recommended to cluster c.
  $relevant(c)$ denotes all items relevant for c.

The error-based evaluation method.
The formula of Mean Absolute Error cluster (MAE_(c)) [13] between the clusters is shown in (5)

$$MAE\_(c) = \frac{\sum_{r(c,j) \in R_c} |r(c,j) - \hat{r}(c,j)|}{|R_c|} \tag{9}$$

The rating of cluster c, $r(c,j)$ is calculated in (12).

$$r(c,j) = \frac{\sum_{i \in c} r(i,j)}{|c|} \tag{10}$$

The prediction of cluster g, $\hat{r}(g,j)$ is calculated in (13).

$$\hat{r}(c,j) = \frac{\sum_{i \in c} \hat{r}(i,j)}{|c|} \tag{11}$$

With c is the number of clusters; R_c shows the set of ratings of cluster c collected the ratings of item; $r(c,j)$ is the real rating of cluster c at the item j; $\hat{r}(c,j)$ is the predicted rating of cluster c at the item j.

3   The Model of the Item-Based Clustering Recommendation

Figure 1 shows a overview of the item-based clustering recommendation model using the energy approach. This recommendation model is described as follows:

− Input (U x I x R)

+ $U = \{u_1, u_2, \ldots, u_n\}$, $u_k \in U$, $k = 1..n$, (including n objects).
+ $I = \{i_1, i_2, \ldots, i_m\}$, $i_j \in I$, $j = 1..m$, (including m attributes).
+ The rating matrix R, with each R_{ui} is a value of R:

$$R_{ui} = \begin{cases} r_{ui} \; if \; the \; user \; u \; rates \; the \; item \; i \\ \emptyset \; if \; the \; user \; u \; dose \; not \; rate \; the \; item \; i \end{cases}$$

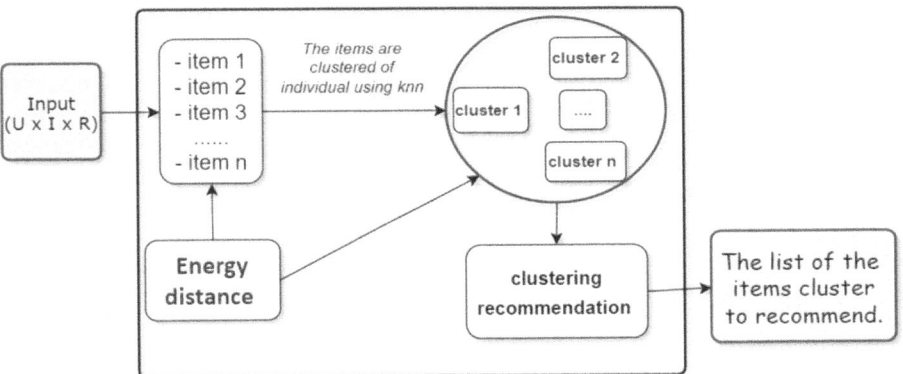

Fig. 1 Item-based clustering recommendation model with energy.

- The energy distance is used to calculate energy between the items and between the clusters. The formula of energy distance are presented in (1) and (5).

- The clustering recommendation is used to predict the missing ratings of the cluster.

- Output is the list of the item clusters with the highest ratings used to recommend Top-N items.

4   Item-Based Clustering Recommendation Algorithm

The energy-based clustering recommendation algorithm is presented in follow:

Algorithm. Energy-based clustering model

Input: The Data Matrix (U x I x R);

Output: Recommending Top-N items;

Begin

[1]: The energy between an item and an item is calculated in I.

   <Matrix[i][j] = Energy $[i_i$ x $i_j]$>;

[2]: The missing rating values of the R matrix is predicted by the knn method using the energy calculated in step 1.

[3]: The items are divided into clusters by using the energy distance.

   <Clustering_List_C[i][j] = Energy [Clustering _C[i]],[Clustering _C[j]]>;

[4]: The ratings of the item cluster are predicted based on the average method of cluster

[5]: The predicted ratings of the item cluster are sorted with DESC.

   < Sort (Clustering_List_C[i])>;

[6]: The n items are recommended with the highest predicted ratings.

   < Print (Top-C[i]>);

End.

---

5 Experiment

5.1 Dataset

The data is selected for the experimental results includes two popular sets, that are: Jester5k dataset [17] and MovieLens Dataset [18]. These two data sets are presented in **Table 2**.

Table 2. Ratings matrix for the item cluster.

Name	Rating	Rating value	Date	Note
Jester5k dataset	5000 x 100 (5000 users and 100 jokes)	From -10 to + 10	April 1999 and May 2003	All chosen users have rated 36 or more jokes
MovieLens Dataset	100,000 (943 users with 1682 movies)	From 1 to 5	Released 4/1998	Each user constraint rated at least 20 movies

5.2   Tool

In the article, the proposed model is built by R language (name EIB - Energy Items_cluster Based is items based clustering collaborative filtering recommendation system using energy distance). This model is compared with two models including: PIB (Pearson Items Based using the Pearson measure) and CIB (Cosine Items Based using the Cosine measure), the PIB and CIB models are available in the "rrecsys" package [14].

Besides, the proposed model is also built in "recommenderlab" package [15] with "energy" package [16] (name IBCF_energy_cluster - Items Based Collaborative Filtering using energy distance to cluster). Compared three models including: IBCF_matching (Item-based collaborative filtering model using matching measure), IBCF_euclidean (Item-based collaborative filtering model using euclidean measure), IBCF_karypis (Item-based collaborative filtering model using the karypis measure), these three models are available in the "recommenderlab" package.

5.3   Scenario 1: Item-Based Clustering by "rrecsys" Package

The experiment result with Jester5k dataset.

This scenario evaluates the error with the MAE_c value of the proposed models EIB (Energy_cluster Items Based using the energy distance) and PIB (Pearson Items Based using the Pearson measure) available in the "rrecsys" package on Jester5k dataset with k nearest items (knn) is 5, 10, 20, 30, 40, 50.

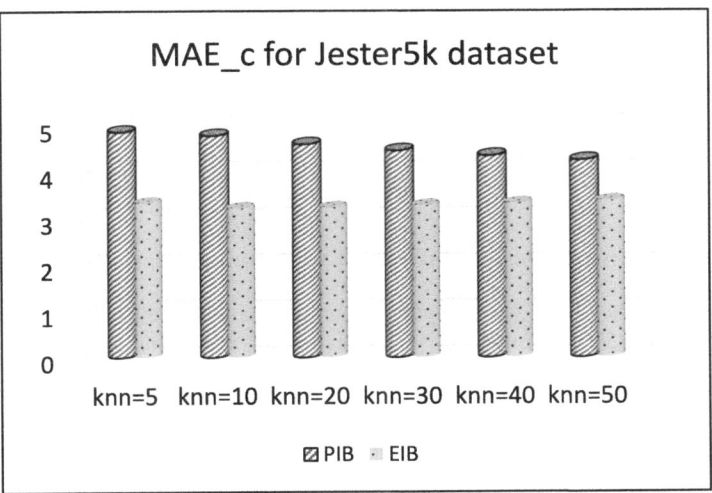

Fig. 2 The MAE_c error value for two models PIB and EIB with Jester5k.

The results in Fig. 2 present: when k nearest items (knn) is 5, 10, 20, 30, 40, 50, the MAE_c error values of EIB are always lower than the MAE_c error values of PIB.

The experiment result with Movielense dataset.

This experiment result showed the MAE_c error value of the proposed models EIB (Energy Items_cluster Based), PIB (Pearson Items Based) and CIB (Cosine Items Based) (the PIB and CIB models are available in the "rrecsys" package) on Movielense dataset with k nearest items (knn) is 30, 40, 50, 60.

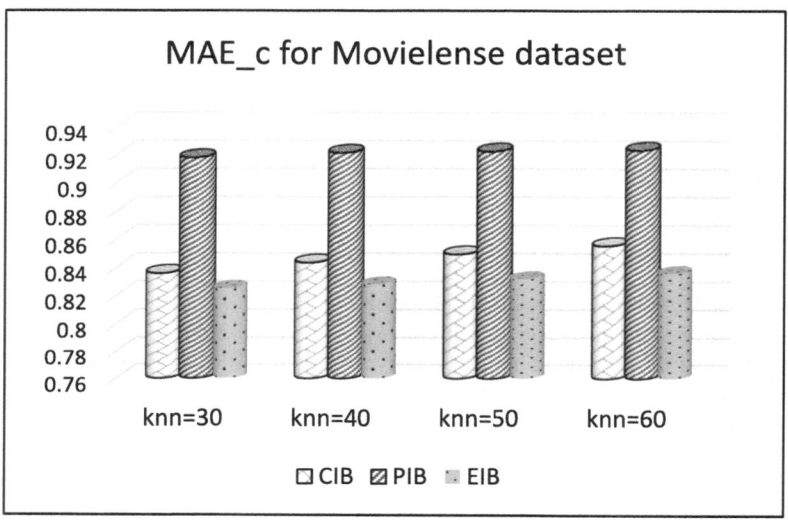

Fig. 3 The MAE_c error for three models EIB, PIB and CIB with Movielense.

Figure 3 Presents that the MAE_c error values of EIB are always lower than the MAE_c error values of PIB and CIB, when k nearest items (knn) is 30, 40, 50, 60.

5.4 Scenario 2: Energy-Based Clustering by "Recommenderlab" Package

This scenario presents item_based clustering recommendation system with energy distance on the Jester5k dataset using two evaluation approach that are the Precision_(c) and Recall_(c). The proposed model (IBCF_energy_cluster) is built with the item-based clustering collaborative filtering model using the energy approach. This model is compared with other models such as: IBCF_matching, IBCF_euclidean, IBCF_karypis, with k nearest items (knn) is 25, 30, 35, 40.

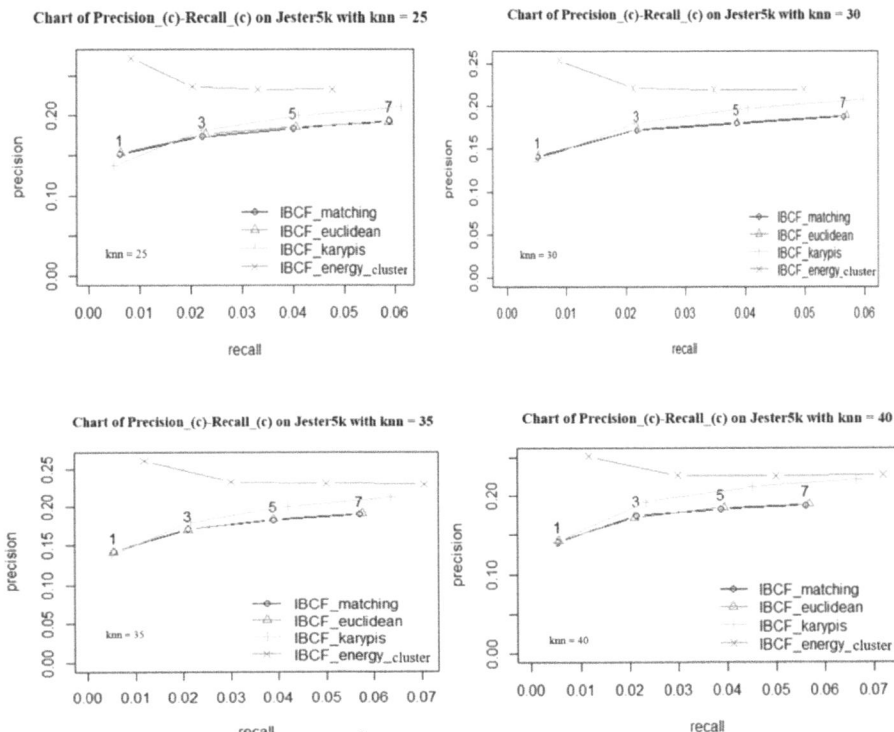

Fig. 4 Figures for chart of Precision_(c) - Recall_(c) with knn = 25, 30, 35, 40 on Jester5k.

Figure 4 presents the Precision_(c) - Recall_(c) values of four models. In which the Precision_(c) - Recall_(c) value of the IBCF_energy_cluster model are always higher than the Precision_(c) - Recall_(c) value of the IBCF_matching, IBCF_euclidean and IBCF_karypis, when k nearest items (knn) is 25, 30, 35, 40.

The experiment result of Fig. 5 presents the ROC curve of four models. In which the ROC curve of the IBCF_energy_cluster model are always higher than the ROC curve of the IBCF_matching, IBCF_euclidean and IBCF_karypis, when k nearest items (knn) is 25, 30, 35, 40.

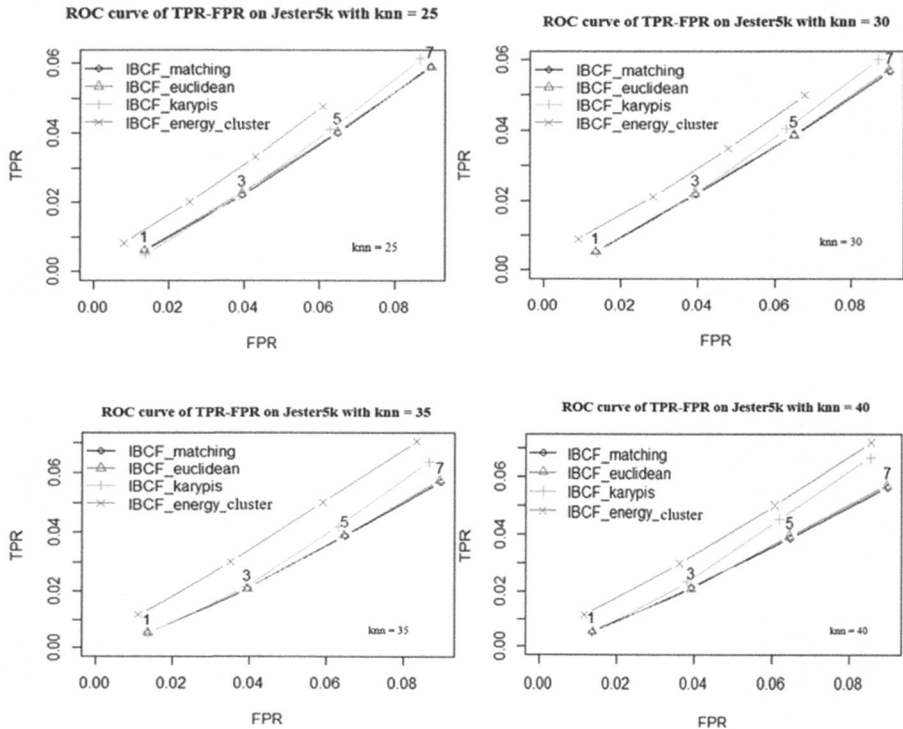

Fig. 5 Figures for ROC curve of TPR-FPR with knn = 25, 30, 35, 40 on Jester5k.

6 Conclusion

In this paper, a clustering algorithm is proposed by us in the item-based collaborative filtering recommendation model using a new energy method to predict the missing ratings of individuals, after predicting the missing ratings of the cluster. Finally, recommend the most relevant items to the user who needs the recommendation based on the predicted rating of the item cluster. The proposed item-based clustering recommendation model was evaluated on both Jester5k and MovieLens by using the MAE_(c) error and the Precision_(c) - Recall_(c) precision value. In general, the cluster proposed model based on the energy distance gives a smaller error than the Peason and Cosine-based comparison model for both datasets in the "rrecsys" package; and the accuracy of the proposed model is higher than the accuracy of the models using the matching, euclidean, and karypis measures in the "recommenderlab" package. Therefore, the item-based clustering recommendation model using the energy distance presents the feasibility of applying potential energy to cluster in the recommendation problems.

References

1. Adomavicius, G., Tuzhilin, A.: Toward the next generation of recommender systems. a survey of the state-of-the-art and possible extensions. IEEE Transactions on Knowledge and Data Engineering, vol 17, 734–749 (2005)
2. Boratto, L B., Carta, S.: State-of-the-art in group recommendation and new approaches for automatic identification of groups", In G. A. Alessandro Soro, Eloisa Vargiu and G. Paddeu, editors, Information Retrieval and Mining in Distributed Environments. Springer Verlag. In press, (2010)
3. Tran, T.C.T., Phan, L.P., Huynh, H.X. (2023). A Group Clustering Recommendation Approach Based on Energy Distance. In: Dinh, T.N., Li, M. (eds) Computational Data and Social Networks . CSoNet 2022. Lecture Notes in Computer Science, vol 13831. Springer, Cham. https://doi.org/10.1007/978-3-031-26303-3_9. (2022)
4. Tran, T, C, T., Phan P. L., Huynh, X, H.: Energy-based collaborative filtering recommendation, International Journal of Advanced Computer Science and Applications(IJACSA), 13(7), 557–562. (2022)
5. Boratto, L B., Carta, S., Satta, M.: Groups Identification and Individual Recommendations in Group Recommendation Algorithms, pp. 27–34. CEUR Workshop Proceedings, (2010)
6. Li, S., Rizzo, L, M.: K-groups: A Generalization of K-means Clustering (2017). ArXiv e-print 1711.04359. https://arxiv.org/abs/1711.04359
7. Li, S.: K-groups: A Generalization of K-means by Energy Distance, Ph.D. thesis, Bowling Green State University (2015)
8. Ntoutsi, I., Stefanidis, K., Norvag, K., Kriegel, HP: gRecs: a Group Recommendation System Based on User Clustering. In: Lee, Sg., Peng, Z., Zhou, X., Moon, YS., Unland, R., Yoo, J. (eds) Database Systems for Advanced Applications. DASFAA 2012. LNCS, vol 7239. Springer, Berlin, Heidelberg (2012)
9. Felfernig, A., Boratto, L., Stettinger, M., Tkalčič, M.: Algorithms for Group Recommendation. In: Group Recommender Systems. SECE, pp. 27–58. Springer, Cham (2018). https://doi.org/10.1007/978-3-319-75067-5_2
10. Dara, S., Chowdary, C.R., Kumar, C.: A survey on group recommender systems. J. Intell. Inf. Syst. **54**, 271–295 (2020)
11. Sarwar, B.M., Karypis, G., Konstan, J. A., Riedl, J.: Item-based collaborative filtering recommendation algorithms. In Proceedings of the 10th international conference on World Wide Web (WWW '01). Association for Computing Machinery, New York, NY, USA, pp. 285–295. (2001)
12. Li, C., Ma, L.: Item-based Collaborative Filtering Algorithm Based on Group Weighted Rating, 2020 13th International Symposium on Computational Intelligence and Design (ISCID), pp. 114–117. Hangzhou, China (2020)
13. Felfernig, A., Boratto, L., Stettinger, M., Tkalčič, M.: Evaluating Group Recommender Systems. In: Group Recommender Systems. SECE, pp. 59–71. Springer, Cham (2018). https://doi.org/10.1007/978-3-319-75067-5_3
14. Çoba, L., Zanker, M., Symeonidis, P.: Environment for Evaluating Recommender Systems, https://rdrr.io/cran/rrecsys/. Repository CRAN, (2019)
15. Hahsler, M.: recommenderlab, A Framework for Developing and Testing Recommendation Algorithm (2015)
16. Rizzo, M., Székely, G.: Energy distance. Wiley Interdisciplinary Reviews: Computational Statistics **8**(1), 27–38 (2016)
17. https://rdrr.io/cran/recommenderlab/man/Jester5k.html. Accessed on 01 Feb 2021
18. https://rdrr.io/cran/recommenderlab/man/MovieLense.html

General Evaluation of EtherCAT-Based Techniques in Various Industrial Systems: Review and Applications

The Tri Bui[1,2], Jin-Ho Shin[3], and Ha Quang Thinh Ngo[1,2(✉)]

[1] Department of Mechatronics, Faculty of Mechanical Engineering, Ho Chi Minh City University of Technology (HCMUT), 268 Ly Thuong Kiet Street, District 10, Ho Chi Minh City, Vietnam
[2] Vietnam National University-Ho Chi Minh City (VNU-HCM), Linh Trung Ward, Thu Duc City, Ho Chi Minh City, Vietnam
[3] Department of Electronic Engineering, Dong-Eui University, Busan, South Korea
nhqthinh@hcmut.edu.vn

Abstract. The need for fieldbus is increasing due to the advancements in industrial control systems. Long transmission distances, high transmission speeds, and reliable real-time performance are crucial requirements for fieldbus technologies. As the demands of the current industry continue to grow, conventional fieldbus solutions struggle to keep up. Consequently, real-time industrial Ethernet technology has become more prevalent in various industries. One of the most widely adopted real-time industrial Ethernet networks is EtherCAT. Its popularity can be attributed to its excellent real-time performance, precise synchronization capabilities, support for numerous topologies, and versatile applicability. This paper aims to provide a valuable scientific reference for researchers engaged in studying EtherCAT.

Keywords: Industrial Network · EtherCAT · Real-time Network · Synchronous Control · Multi-axes Motion Control

1 Introduction

With standard physical connections and increasing speeds, Ethernet has gained widespread adoption. Industrial Ethernet, a development of Ethernet technology, involves modifications to the Media Access Control (MAC) layer. This technology offers compatibility, cost-effectiveness, high bandwidth, and flexible topology, which has led to its gradual replacement of traditional Fieldbus technology. As a result, many industrial communication protocols are transitioning to industrial Ethernet-based solutions. In [1], researchers highlight five primary industrial real-time Ethernet networks, i.e. Powerlink [2], PROFINET [3], SERCOS III [4], Ethernet/IP [5], Real-time Ethernet/IP [6] and EtherCAT [7]. Among these protocols, EtherCAT has garnered significant attention from researchers due to its determinism and real-time control capabilities.

© ICST Institute for Computer Sciences, Social Informatics and Telecommunications Engineering 2024
Published by Springer Nature Switzerland AG 2024. All Rights Reserved
P. Cong Vinh and N. Thanh Tung (Eds.): ICCASA 2023, LNICST 579, pp. 126–138, 2024.
https://doi.org/10.1007/978-3-031-58878-5_9

EtherCAT, developed by Beckhoff Automation, is a real-time Ethernet network that is gaining popularity in factory automation environments [8]. While it is based on conventional Ethernet technology, EtherCAT utilizes a unique approach to access slave devices. It also defines a logical addressing scheme that allows for efficient packing of small process data, resulting in high communication efficiency and very short cycle times. This efficiency makes it an attractive solution for connecting peripherals like I/O devices and servo drives to the application master. In addition to its efficiency, Ether-CAT offers another compelling feature that appeals to various application domains, especially motion control. This feature is the ability to enable synchronous operation of up to 65535 devices through a simple yet effective mechanism. The Distributed Clock (DC) technology of EtherCAT [9] ensures that local clocks in all EtherCAT devices run in synchronization. This capability is increasingly essential in control systems and adds to the appeal of EtherCAT in different industrial applications.

Given the widespread usage of EtherCAT and the numerous studies conducted on this technology in various aspects, it is indeed an opportune moment to review the state of the art in this area. Our review should focus on studies related to master and slave systems [10], the synchronization mechanism [11], and its applications [12]. Conducting a comprehensive literature review is considered an appropriate approach to identify and understand modern approaches towards EtherCAT and to explore its capabilities fully. This investigation summarizes and analyzes a large scale of theoretical researches on EtherCAT, which contain main discussions of the collected data. The rest of our works is as follows. Section 2 synthesizes the hardware and software of EtherCAT master. Then, Sect. 3 investigates the utility of various EtherCAT slave controllers. Subsequently, the algorithms and methods for synchronization mechanism are explored in Sect. 4. Some applications of EtherCAT technologies are denoted in Sect. 5. Finally, conclusions and future works are mentioned in Sect. 6.

2   EtherCAT Master Station

The EtherCAT master research can be categorized into two main directions: hardware and software. On the software side, EtherCAT is an open Ethernet protocol adhering to global standards, which has led to the development of numerous commercial solutions and open-source projects. Notably, open-source real-time Linux platforms like RT-Preempt, RTAI, and Xenomai are widely utilized in combination with the EtherCAT master stack provided by IgH, as well as with open-source EtherCAT master stacks like Simple Open EtherCAT Master (SOEM) from the Open EtherCAT Society. The adoption of an open-source-based EtherCAT master system offers several advantages, including flexibility in application program development and cost reduction. Thanks to the open-source nature of these platforms, developers can create applications in the C/C ++ programming language, without being restricted by certain limitations imposed by proprietary solutions like Beckhoff's TwinCAT. For instance, TwinCAT imposes restrictions on memory allocation and the use of standard C/C ++ math functions, making it challenging to develop applications with complex features, such as dynamics-based manipulator control. In contrast, open-source real-time Linux platforms provide both kernel space and user space interfaces.

Table 1. Summary of the state-of-the-art for related researches in EtherCAT master

Author(s)	Publication year	Main research	Concern	Operating System	EtherCAT-based master software	Advantage(s)	Disadvantage(s)
Shi, H. et al. [1]	2022	Hardware	EtherCAT hardware based on development board	N/A	N/A	A hardware EtherCAT master is introduced based on the ARM architecture and has a PCIe port to connect to the computer. Hardware guarantees about Hard real-time	The hardware structure is quite complex and the connection to the computer is still underdeveloped
Yi, H. C. et al. [13]	2019		Cycle time improvement	Linux with RT-Preempt	Igh Etherlab	Presents a hardware architecture under the linux kernel that does not use standard drivers for all hardware, but is developed specifically for a certain network hardware. This reduces system time and delay by eliminating memory copy and NAPI	Just show the general structure, not detail the structure of the developed Direct Ethernet Drive
Zhang, H. et al. [14]	2019		EtherCAT hardware based on development board	N/A	N/A	EtherCAT master is developed based on FPGA chip, with this structure can be developed for other circuits	The experiment has not yet clearly stated the performance of the hardware, but only tested a certain system cycle
Kim, S. et al. [15]	2017	Software	Performance analysis of open-source software	Linux with RT-Preempt, RTAI, Xenomai	Open Source-based EtherCAT Master (SOEM, Igh Etherlab)	The article shows the experiments of open-source libraries on different Linux-based real-time operating systems and compares them with TwinCAT, thereby showing the low-delay capabilities of the Linux architecture	Only shows the results of the experiments, but there are no graphs to demonstrate the system's capabilities
Shi, B. H. et al. [16]	2017		Software development based on open-source software	Linux with RTAI	Igh Etherlab	Use Igh EtherCAT library to add LinuxCNC software to control EtherCAT servo drivers for CNC control purposes	The experimental section only mentions the software and setup, not if the experimental results, figures or graphs

(*continued*)

Table 1. (*continued*)

Author(s)	Publication year	Main research	Concern	Operating System	EtherCAT-based master software	Advantage(s)	Disadvantage(s)
Alex, B. et al. [17]	2021		Software development based on open-source software	Linux without RT patch	SOEM	Develop SOEM-based extension libraries for EtherCAT control, including fastcat and jsd. Libraries provide the ability to connect variables of functions and slaves together using only YAML config files. Guaranteed zero delay even without using real-time kernel	Support a small number of devices, if you want to add the devices you want, you need to have knowledge of that hardware
Cho, S. Y. et al. [18]	2023		Software development based on open-source software	Linux with RT-patch	Igh Etherlab	Develop an extension for Python programming on real-time Linux environment based on Igh Etherlab. Experimental comparison of delay is not too different when using C language when programming real-time tasks	The system cycle is limited to a minimum of 1ms, while currently aiming for smaller send and receive cycles

N/A: Not applied

In terms of hardware, developers are often driven to develop solutions that aim to reduce latency and enhance the processing speed of the EtherCAT master. The effectiveness of control systems heavily relies on an EtherCAT master with low latency and high synchronization capabilities, enabling fast and precise execution of actions. However, as high-end control fields advance, conventional EtherCAT masters face limitations in meeting the performance demands of such advanced control systems, mainly due to real-time constraints. Therefore, the development of a new EtherCAT master with improved real-time performance becomes crucial to enable its effective utilization in cutting-edge industrial applications. Cycle time and clock synchronization are vital benchmarks for evaluating EtherCAT's real-time performance. Achieving optimal cycle times and precise clock synchronization is essential to ensure efficient operation and coordination of industrial processes. Table 1 presents various studies focusing on the EtherCAT master, along with their respective advantages and limitations. These studies aim to address the challenges posed by real-time performance in EtherCAT systems and strive to enhance the technology's capabilities for advanced industrial applications. By analyzing these studies, researchers can gain valuable insights into the progress made in EtherCAT master development and identify potential areas for further improvement.

3   EtherCAT Slave Station

An EtherCAT slave system comprises the Physical layer, Data Link layer, and Application layer. The architecture of the EtherCAT slave is depicted in Fig. 1. The Physical layer includes essential physical components such as RJ45 connections, magnetics, and standard PHYs, which are responsible for processing fieldbus signals [19, 20]. By adhering to the Ethernet standard IEEE 802.3, the Physical layer transmits network data to the EtherCAT slave controller (ESC) and applies signals from the ESC to the network.

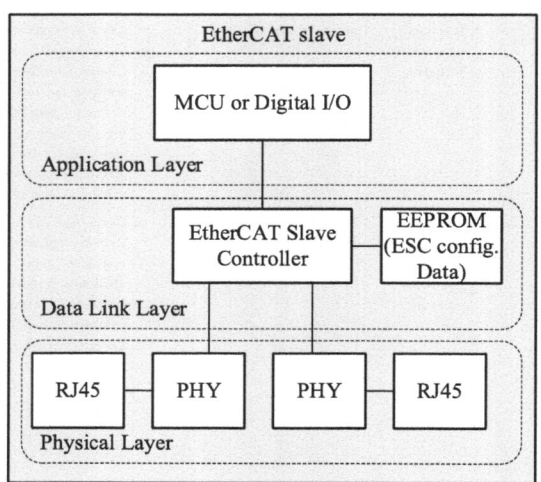

Fig. 1. Architecture of the EtherCAT slave station

Moving to the Data Link layer, both the ESC and EEPROM store the initial configuration. The ESC serves as the primary communication processor for the slave, and it stores the frame data processed in the Dual-port RAM (DPRAM). When dealing with a straightforward application, the application layer can be implemented on the digital I/O interface. However, for more complex processing requirements, the application layer is implemented on the local host MCU (Microcontroller Unit). This setup allows for more sophisticated data processing and control capabilities. By understanding the architecture and components of an EtherCAT slave system, engineers can effectively design and implement EtherCAT-based solutions to meet the specific needs of industrial automation and control applications.

EtherCAT slave controllers are typically built on either an ASIC (Application-Specific Integrated Circuit) or FPGA (Field-Programmable Gate Array) platform. These ESCs can be classified based on various factors, including type, manufacturer, package type, size, DPRAM capacity, FMMU (Fieldbus Memory Management Unit) support, distributed clock capability, and other relevant features, depending on the specific requirements of the research or application. Table 2 provides an overview of various controllers available in selected studies, offering valuable insights into the different options researchers have explored in the realm of EtherCAT slave controllers. This information is beneficial for understanding the existing ESC landscape and for making informed decisions when selecting the appropriate controller for a particular EtherCAT-based project.

Table 2. List of the state-of-the-art for related researches in EtherCAT slave

Author(s)	Publication year	EtherCAT-based slave controller	μC Interface	Main μC	Sync signal	Application
Nguyen, V. Q. et al. [21]	2017	ET100	SPI	TMS320F2812	N/A	Closed-Loop Stepper Motor Drive
Liu, J. et al. [22]	2020	ET100	Parallel 16-bit	Microprocessor Zynq-7020	x	I/O module
Fey, J. H. et al. [23]	2019	FPGA with EtherCAT IP Core	Parallel 16-bit	FPGA	x	Modular Multilevel Converter
Zheng, L. et al. [24]	2020	AX58100	SPI	STM32G431	N/A	Step servo
Jiao, B. et al. [25]	2014	ESC20	Parallel 16-bit	TMS320F2812 DSP	x	Steel Plate Loading and Unloading System
Jing, H. et al. [26]	2023	LAN9252	Parallel 16-bit	STM32F767	x	DC Servo driver
Herron, C. et al. [27]	2023	LAN9252	SPI	TM4C123GXL	x	DC Servo driver
Mishra, H. et al. [28]	2022	CIFX90E-RE	mPCIe	SBC MIG 5251	N/A	HMI module

4   EtherCAT-Based Synchronization Mechanism

The EtherCAT DC allows all EtherCAT devices to share a system time for managing device node synchronization. Accurate synchronization is known to be crucial to the synchronization process. For instance, in a CNC system, multiple joint points must operate concurrently and move in together. Serious repercussions will result if there is a significant difference between the nodes during the synchronization period. The accuracy of the EtherCAT clock synchronization algorithm needs to be improved, despite having reached a certain level.

On the EtherCAT bus with DC, there are three types of clocks: local clock, master clock and reference clock. Each slave has an internal clock known as a local clock, which presents the time that begins at zero when a slave is turned on. The master clock is the clock of EtherCAT master, which is usually an IPC with Windows or Linux OS. Therefore, compared to an EtherCAT slave clock, the precision of the master clock is often lower, and jitters are higher. The first DC-capable EtherCAT slave's local clock serves as the reference clock. Because the DC Master will read the reference clock and distribute it to the other slaves in the same frame, the first slave is commonly selected to become a reference slave.

The latency or time difference in the EtherCAT bus comes from three sources: propagation delay, offset, and drift. All slaves must be reached before the EtherCAT frame returns to the master. When a data frame enters a slave, the slave also needs a small amount of time to process the data before moving on to the next slave. All of these delays are known as propagation delays. Propagation delay measurement and offset compensation can be done once at the starting up and during the Pre-OP state. The clock drift can also be dynamically adjusted by using the master's data frame.

The clock synchronization algorithm of EtherCAT contains two main concerns, master - reference slave synchronization and reference slave–slave synchronization. Table 3 discusses the clock synchronization algorithm studies, including methods, advantages and disadvantages.

5   Application of EtherCAT

In the past decade, EtherCAT technology has gained popularity in robot applications [35, 36] and multi-axis systems [37, 38] mostly because of its adaptability, substantially decreased wiring, and low cycle time. In this article, representative examples from the various robotic application domains are provided below and in Table 4.

CW Hung et al. [39] proposed the delta robot control system based on EtherCAT for painting applications in the robot field. The experimental results show that the Cyclic Sync Velocity mode with PD controller is better accuracy when painting than the Cyclic Sync Position mode. However, it is necessary to adjust the proportional coefficient and derivative coefficient of the motor drivers to prove the accuracy.

Table 3. Summary of the state-of-the-art for related researches in synchronization mechanism

Author(s)	Publication year	Concern	OS	Methods	Advantage	Disadvantage
Chen, X. et al. [29]	2016	Synchronization between master and reference slave	Windows with real-time extension (RTX)	The algorithm changes the frequency of the master clock every cycle by taking the master clock value from the previous cycle and calculating it with the cycle of the reference slave	Offering an option of controlling and synchronizing masters with EtherCAT slaves on real-time Windows is RTX. The synchronous method requires no setup time at runtime	Experiments do not clearly show the master - reference slave synchronization of the algorithm, but only demonstrate system cycle has low jitter when using RTX

(continued)

Table 3. (*continued*)

Author(s)	Publication year	Concern	OS	Methods	Advantage	Disadvantage
Park, S. M. et al. [30]	2020		Linux with RT patch	Firstly, synchronize the master - reference slave in the init state to calculate and give the offset time to the reference clock. When in the operation state, the master will take the drift time of the reference slave and use the EMA filter to calculate and shift the system time for other slaves	This article presents a method of synchronizing the clocks by writing to the system time register of all the slaves in the network with the clock shift value. This doesn't require master time shifting	The system requires a long setup time to stabilize each time it works
Park, S. M. et al. [31]	2021		Linux with RT patch	Using GPIO signals to measure the GPI of the master and the time when the external processor starts sending values to shared memory to compensate for the synchronization delay on the maste	Provides a synchronization method for external processors - EtherCAT master shares memory and ensures that data will not be lost	The number of samples is quite large, so even in init or operation state, it takes a long time for the system to stabilize
Libo, C. et al. [32]	2023		Non-OS	Use the master and reference slave cycle times and then apply the PD algorithm to predict the next cycle shift of the master	Given the synchronous algorithm used in ARM microcontrollers with not too high frequency and no OS in embedded CNCs	The algorithm has time values that are not taken from the operation state but calculated by formulas, which leads to sometimes different calculated and actual values, leading to the system output may be wrong

Table 4. Summary of the state-of-the-art for related researches in EtherCAT applications

Author(s)	Publication year	Platform	Robot configuration	Number of slaves	Cycle time	HMI	Control Algorithm	Real-time OS included
Hung, C. W. et al. [39]	2022	Delta robot	A type of parallel robot that consists of three arms connected to universal joints at the base. The key design feature is the use of parallelograms in the arms, which maintains the orientation of the end effector	3	4	x	x	x
Delgado, R. et al. [40]	2016	Omnidirectional Mobile Robot	A simple omnidirectional mobile robot using 4 mecanum wheels. Common mecanum wheel mobile robots are either in the form of a parallelogram. The mecanum wheels are attached to the robot system to constitute a basic 45° driving system	4	1	N/A	x	x
Jia, H. et al. [41]	2017	CNC	A Wear-resistant coating testing system based on the CO-TRUST C37 motion controller. Connected via EtherCAT bus, the CO-TRUST C37 motion controller and the servo system are mainly responsible for the motion control of all the axes of the wear-resistant coating testing equipment, executing the motion code, determining the logic function of the system PLC and communicating with the host computer	6	N/A	x	x	N/A
Ahn, J. W. et al. [7]	2023	Humanoid Robot	The Humanoid Robot TOCABI was designed to reflect the height of an adult male, whose weight and height are 100 kg and 1.8 m, respectively. TOCABI has 33 joints, with two DOFs in the neck, eight in each arm, three in the waist, and six in each leg	33	0.2–0.25	x	N/A	x

(continued)

Table 4. (*continued*)

Author(s)	Publication year	Platform	Robot configuration	Number of slaves	Cycle time	HMI	Control Algorithm	Real-time OS included
Yuan, L. et al. [37]	2022	Multi-axis high-precision positioncontrol in material transportation	The system is a multi-axis position control system composed of four drive servo mechanisms of storing-storage, storing-supply, stop-material and supply	4	2.5	x	x	N/A
Zhang, G. et al. [42]	2019	7-DoF light-weight cooperative robot	A 7-DoF cooperative manipulator based on EtherCAT bus only needs 4 cables to communicate with the PC, and 2 cables to obtain power	1	1	N/A	N/A	x

N/A: Not applied

For mobile robot applications, the investigation in [40] developed an EtherCAT-based four-wheel omnidirectional mobile robot using mecanum wheels. The results of this experiment demonstrate that, with a minimum amount of jitter and an acceptable execution time, an open-source EtherCAT Master can function as the main controller of a mobile robot control system. In [41], there is an example of a CNC application. The CO-TRUST C37 motion controller was programmed by the authors using CODESYS software. The hardware platform employed an EtherCAT bus. The CNC controller supports I/O devices through the EtherCAT interface in addition to servo drivers.

The ability to control many axes is shown in the research [7] which presents a dual-channel real-time EtherCAT control system for the 33 degrees of freedom (DOF) TOCABI humanoid robot. The performance validation showed how to set up the dual-channel EtherCAT MainDevice so that it can drive robots with a lot of degrees of freedom at a faster communication cycle.

Some researchers [37] present the application of EtherCAT technology in material transportation to solve the problem of multi-axis high-precision position control PLC master controller is used as a EtherCAT master with integrated input and output module, power failure hold module. Furthermore, another development [42] use the EtherCAT bus as communication between the 7-DoF lightweight robot and ROS controller. The SOEM library with the real-time kernel Xenomai controls the robot when it receives the command from the ROS. The results show that the real-time control cycle is stable.

6 Conclusions

To the best of our knowledge, EtherCAT has the potential to be an applicable protocol for all industrial and manufacturing applications, thanks to its advanced technology and superior performance. This paper provides a review of researches in the past decade on specific aspects of EtherCAT that have garnered a lot of attentions. The analysis of EtherCAT master focuses on two key issues: hardware and software. Additionally, different types of ESCs and their respective uses are compared. Furthermore, the advantages, disadvantages, and methods of each synchronization algorithm are explained and analyzed. Finally, several applications through illustrative examples are presented.

Acknowledgement. We acknowledge Ho Chi Minh City University of Technology (HCMUT), VNU-HCM for supporting this study.

References

1. Shi, H., Lin, W., Liu, C., Jinyong, Yu.: A novel heterogeneous parallel system architecture based EtherCAT hard real-time master in high performance control system. Electronics **11**(19), 3124 (2022). https://doi.org/10.3390/electronics11193124
2. Romanov, A., Slepynina, E.: Real-time Ethernet POWERLINK communication for ROS. Part I. General concept. In: 2020 Ural Smart Energy Conference (USEC), pp. 159–162. IEEE (2020)
3. Turcato, A.C., Negri, L.H.B.L., Dias, A.L., Sestito, G.S., Flauzino, R.A.: A cloud-based method for detecting intrusions in profinet communication networks based on anomaly detection. J. Control Autom. Elect. Syst. **32**(5), 1177–1188 (2021)
4. Sestito, G.S., Turcato, A.C., Dias, A.L., Ferrari, P., da Silva, M.M.: Evaluating Real-Time Ethernet performance indicators for SERCOS III networks. In 2021 14th IEEE International Conference on Industry Applications (INDUSCON) (pp. 1191–1197). IEEE (2021).
5. Lindner, S., Häberle, M., Menth, M.: P4TG: 1 Tb/s Traffic Generation for Ethernet/IP Networks. IEEE Access **11**, 17525–17535 (2023)
6. Truong, Q. V., Thinh Ngo, H. Q.: Control and Implementation of Positioning System with Symmetrical Topology for Precision Manufacturing. Mathematical Problems in Engineering, (2022)
7. Ahn, J., Park, S., Sim, J., Park, J.: Dual Channel EtherCAT Control System for 33-DOF Humanoid Robot TOCABI. IEEE Access (2023)
8. Jansen, D., Buttner, H.: Real-time Ethernet: the EtherCAT solution. Comput. Control. Eng. **15**(1), 16–21 (2004)
9. Shen, H., Li, P., Luo, X.: Synchronous multi-axis motion control based on modified Ether-CAT distributed clock. In 2020 Chinese Automation Congress (CAC), pp. 3674–3678. IEEE (2020).
10. Song, G., & Lei, J.: Control system design of heterogeneous master-slave robot for fracture reduction surgery of long bone. In 2023 IEEE International Conference on Mechatronics and Automation (ICMA) (pp. 2303–2308). IEEE (2023)
11. Paprocki, M., Erwiński, K.: Synchronization of electrical drives via EtherCAT fieldbus communication modules. Energies **15**(2), 604 (2022)
12. Nguyen, T.P., Nguyen, H., Ngo, H.Q.T.: Developing and Evaluating the Context-Aware Performance of Synchronization Control in the Real-Time Network Protocol for the Connected Vehicle. Mobile Netw. Appl. (2023). https://doi.org/10.1007/s11036-023-02182-y
13. Yi, H.C., Choi, J.Y.: Cycle time improvement of EtherCAT networks with embedded linux-based master. IEICE Trans. Inf. Syst. **102**(1), 195–197 (2019)
14. Huawei Zhang, Y., Xiang, H.S., Qin, F., Niu, Z.: Design and Implementation of Ethercat Master Based On ZYNQ. IOP Conf. Ser: Mater. Sci. Eng. **612**(4), 042052 (2019). https://doi.org/10.1088/1757-899X/612/4/042052
15. Kim, S., Shin, E.: A performance evaluation of open source-based EtherCAT master systems. In: Proceedings 4th International Conference Control, Dynamics, Systems, Robot, pp. 128–1 (2017).

16. Bu-Hai, S., Yong-Zhi, W., Chuan, D.: A design of realtime communication based on EtherCAT in industrial robot control system based on LinuxCNC. In 2017 29th Chinese Control And Decision Conference (CCDC) (pp. 5776–5780). IEEE (2017).

17. Brinkman, A., Morris, J., Chen, I., Sheikh, N., Warren, P.: Fastcat: an open-source library for composable EtherCAT control systems. In 2021 IEEE Aerospace Conference (50100), pp. 1–8. IEEE (2021).

18. Cho, S.Y., Delgado, R., Choi, B.W.: Feasibility Study for a Python-Based Embedded Real-Time Control System. Electronics 12(6), 1426 (2023)

19. Park, S.M., Kim, H.W., Cho, H.M., Choi, J.Y.: Development of EtherCAT slave based on multi-core DSP. In 2018 15th International Conference on Control, Automation, Robotics and Vision (ICARCV), pp. 157–161. IEEE (2018)

20. Nguyen, Q.V., Kim, T.W., Moon, J.Y., Jeon, J.W.: Development of independent EtherCAT slave module and application to closed loop step motor drive with multi-axis. In 2016 International Conference on Computing, Communication and Automation (ICCCA) (pp. 912–917). IEEE (2016).

21. Nguyen, V.Q., Tran, N.V.P., Tran, H.N., Le, K.M., Jeon, J.W.: A closed-loop stepper motor drive based on EtherCAT. In IECON 2017–43rd Annual Conference of the IEEE Industrial Electronics Society (pp. 3361–3365). IEEE (2017)

22. Liu, J., Zhang, H., Guo, X., Chen, W.: Design of ethercat slave system based on zynq-7020 chip. In 2020 15th IEEE Conference on Industrial Electronics and Applications (ICIEA) (pp. 1916–1920). IEEE (2020)

23. Fey, J.H., Hinrichsen, F., Carstens, G., Mallwitz, R.: Development of a modular multi-level converter demonstrator with EtherCAT communication. In 2019 IEEE 13th International Conference on Compatibility, Power Electronics and Power Engineering (CPE-POWERENG) (pp. 1–6). IEEE (2019)

24. Zheng, L., Zhangyu, L., Liu, Z., Tan, C.: Design of Step Servo Slave System Based on EtherCAT. In: Liu, Q., Liu, X., Li, L., Zhou, H., Zhao, H.H. (eds.) Proceedings of the 9th International Conference on Computer Engineering and Networks. AISC, vol. 1143, pp. 193–205. Springer, Singapore (2021). https://doi.org/10.1007/978-981-15-3753-0_19

25. Jiao, B., He, X.: Application of the real-time EtherCAT in steel plate loading and unloading system. In: Li, K., Xue, Y., Cui, S., Niu, Q. (eds.) Intelligent Computing in Smart Grid and Electrical Vehicles, pp. 268–275. Springer Berlin Heidelberg, Berlin, Heidelberg (2014). https://doi.org/10.1007/978-3-662-45286-8_28

26. Jing, H., Chen, W., Bai, S., Bai, Y.: EtherCAT industrial ethernet slave design and application study. In Second International Symposium on Computer Applications and Information Systems (ISCAIS 2023) (Vol. 12721, pp. 198–204). SPIE (2023)

27. Herron, C.W., Fuge, Z.J., Kogelis, M., Tremaroli, N.J., Kalita, B., Leonessa, A.: Design and validation of a low-level controller for hierarchically controlled exoskeletons. Sensors 23(2), 1014 (2023)

28. Mishra, H., Saini, L. M., Bhandwale, A.: Design of EtherCAT Slave Controller using CIFX 90E-RE for HMI Display. In 2022 International Conference on Connected Systems & Intelligence (CSI) (pp. 1–6). IEEE (2022)

29. Chen, X., Li, D., Wan, J., Zhou, N.: A clock synchronization method for EtherCAT master. Microprocess. Microsyst. 46, 211–218 (2016)

30. Park, S.M., Kim, H.W., Kim, H.J., Choi, J.Y.: Accuracy improvement of master–slave synchronization in EtherCAT networks. IEEE Access 8, 58620–58628 (2020)

31. Park, S.M., Kwon, Y., Choi, J.Y.: Time synchronization between EtherCAT network and external processor. IEEE Commun. Lett. 25(1), 103–107 (2020)

32. Libo, C., Taiyong, W., Songhui, J., Chong, T., Ying, T.: Innovation of EtherCAT adaptive synchronization control in embedded CNC. Int. J. Commun. Syst. 36(8), e5462 (2023)

33. Liu, J., Yang, L., Xu, D., Wu, X.: A high precision clock synchronization algorithm for the EtherCAT. In 2017 12th IEEE Conference on Industrial Electronics and Applications (ICIEA) (pp. 1369–1374). IEEE (2017).
34. Park, S.M., Kim, H., Kim, H.W., Cho, C.N., Choi, J.Y.: Synchronization improvement of distributed clocks in EtherCAT networks. IEEE Commun. Lett. **21**(6), 1277–1280 (2017)
35. Phan, D.Q., Ngo, H.Q.T.: Implementation of multiple controllers for context-inspired collaboration between human and robot by integrating the uni-axial motion and real-time operating system. Internet of Things **22**, 100788 (2023)
36. Sygulla, F., et al.: An EtherCAT-based real-time control system architecture for humanoid robots. In 2018 IEEE 14th International Conference on Automation Science and Engineering (CASE) (pp. 483–490). IEEE (2018)
37. Yuan, L., Guan, X., Guan, S., Baoqi, W.: Design of Multi-axis Motion Control System Based on EtherCAT. In: Yang, Q., Li, J., Xie, K., Jianlin, H. (eds.) The Proceedings of the 17th Annual Conference of China Electrotechnical Society: Volume I, pp. 321–332. Springer Nature Singapore, Singapore (2023). https://doi.org/10.1007/978-981-99-0357-3_33
38. Nguyen, H., Nguyen, T.P., Ngo, H.Q.T.: Improving the tracking performance under nonlinear time-varying constraints in motion control applications: from theoretical servo model to experimental validation. Math. Probl. Eng. **2021**, 1–15 (2021)
39. Hung, C.W., Tseng, Y.H., Jiang, G.Y., Song, C.C.: An EtherCAT based delta robot synchronous control application. J. Robot. Netw. Artif. Life **9**(2), 183–186 (2022)
40. Delgado, R., Shin, W.C., Hong, C.H., Choi, B.W.: Development and control of an omnidirectional mobile robot on an ethercat network. Int. J. Appl. Eng. Res. **11**(21), 10586–10592 (2016)
41. Jia, H., Yao, P., Li, B., Tian, X.: Four axes wear-resistant coating testing system based on EtherCAT. In 2017 Chinese Automation Congress (CAC) (pp. 2842–2846). IEEE (2017)
42. Zhang, G., Li, Z., Ni, F., Liu, H.: A real-time robot control framework using ROS control for 7-DoF light-weight robot. In 2019 IEEE/ASME International Conference on Advanced Intelligent Mechatronics (AIM) (pp. 1574–1579). IEEE (2019).

Towards an IoT-Based Unmanned Surface Vehicle Design for Environment Monitoring in Mekong Delta

Cuong Pham-Quoc[1,2]([envelope]) [iD] and Nguyen Cao Tri[1,2]

[1] Ho Chi Minh City University of Technology (HCMUT), 268 Ly Thuong Kiet Street, District 10, Ho Chi Minh City, Vietnam
[2] Vietnam National University- Ho Chi Minh City (VNU-HCM), Ho Chi Minh City, Vietnam
cuongpham@hcmut.edu.vn

Abstract. Water quality is a major environmental issue and one of humanity's major issues. For example, although the Mekong Delta has massive water resources from lakes, rivers, and aquifers, the area suffers from problems due to the reduced usable water supplies. Water pollution and salinization have become critical issues in most nations worldwide due to oil spills, plastic waste, sea-level increase, and human activities. Contamination of this nature can harm fish and other aquatic life habitats, agriculture, and, eventually, human health. This paper introduces our IoT-based Unmanned Surface Vehicle design for monitoring the Mekong Delta wetland environment. We explore the use of recent advances in open-source Global Positioning System (GPS)-guided drone technology to design and test a low-cost and transportable small unmanned surface vehicle (sUSV). The vehicle operates using Ardupilot open-source software and can be used by local scientists and marine managers to map and monitor marine environments in shallow areas with commensurate visibility. The USV is equipped with multiple sensors for measuring various water's parameters at different positions. The experimental results show that the prototype version of our USV can work 573 m away from the base station while mean square error (MSE) of telemetry data from sensors compared to certified handheld devices is only 6.2%.

1 Introduction

Unmanned Surface Vehicles (USVs) offer several advantages for wetland environmental management. One of the key advantages is the remote Monitoring where USVs can be equipped with various sensors and monitoring devices to collect real-time data about wetland ecosystems. This includes water quality parameters such as temperature, pH levels, dissolved oxygen, and nutrient concentrations. Remote monitoring eliminates the need for manual data collection and allows for continuous, high-resolution monitoring of wetland conditions. USVs

P. Cong Vinh and N. Thanh Tung (Eds.): ICCASA 2023, LNICST 579, pp. 139–148, 2024.
https://doi.org/10.1007/978-3-031-58878-5_10

offer the following advantages for environmental management: (i) Reduced Ecological Impact: Unlike manned vessels, USVs do not require human operators on board, minimizing the physical disturbance to sensitive wetland habitats. (ii) Versatility and Adaptability: USVs can be easily programmed and reconfigured for different tasks and environments. They can navigate narrow channels, traverse complex terrain, and access hard-to-reach areas of wetlands. (iii) Cost-Effectiveness: Operating USVs for wetland environmental management can be more cost-effective compared to manned vessels or aerial platforms. They require fewer resources, such as fuel and maintenance, and can be deployed for extended periods without the need for human presence. (iv) Improved Safety: Wetlands can present hazardous conditions for human operators due to unpredictable terrain, dense vegetation, or exposure to toxins. By utilizing USVs, the risk to human life and health is significantly reduced as they eliminate the need for personnel to physically enter potentially dangerous areas. (v) Enhanced Data Accuracy and Integration: USVs can be equipped with advanced sensors and data acquisition systems that provide accurate and reliable measurements. The collected data can be integrated with Geographic Information Systems (GIS) and other analytical tools, allowing for comprehensive data analysis and visualization. (vi) Reduced Environmental Footprint: USVs are typically electrically powered, which reduces their carbon footprint and eliminates direct emissions. They can utilize renewable energy sources, such as solar panels or hydrokinetic energy, for prolonged operation in remote wetland areas.

While Unmanned Surface Vehicles (USVs) offer numerous benefits for environmental management, there are also some technology-related challenges that need to be addressed. Here are a few notable issues: (i) Navigation and Obstacle Avoidance: USVs must navigate through complex environments, including waterways with varying water levels, submerged obstacles, and vegetation. (ii) Autonomous Decision-Making: USVs often operate autonomously or with limited human intervention. (iii) Communication and Connectivity: USVs often rely on communication links, such as satellite or cellular networks, to transmit data, receive commands, or maintain remote control. (iv) Power Supply and Endurance: USVs typically rely on onboard batteries or alternative power sources, such as solar or wind energy, to operate. (v) Sensor Integration and Data Fusion: USVs often carry multiple sensors and data acquisition systems to monitor environmental parameters. (vi) Data Processing and Analysis: USVs generate vast amounts of data during environmental monitoring missions. (vii) System Reliability and Redundancy: USVs are expected to operate reliably over extended periods, often in harsh environmental conditions. Addressing these technology issues will contribute to the continued advancement and successful deployment of USVs for environmental management, improving their reliability, performance, and overall contribution to the preservation and sustainable management of natural ecosystems.

The main purposes and contributions of this paper is to present a novel design and implementation of an IoT-based Unmanned Surface Vehicle (IoT-USV) tailored explicitly for environmental monitoring applications. The proposed IoT-USV system aims to provide the following functionalities:

– Autonomous navigation: The USV will be equipped with advanced navigation algorithms and obstacle avoidance mechanisms to operate efficiently in dynamic aquatic environments.
– Multi-sensor data collection: Integrating a variety of sensors, such as water quality analyzers, temperature sensors, GPS, and cameras, to acquire comprehensive and real-time environmental data.
– Data transmission and communication: Establishing a reliable and secure IoT infrastructure to facilitate seamless data transmission from the USV to an onshore base station for immediate analysis and decision-making. Energy-efficient operation: Optimizing power consumption to ensure prolonged mission durations and reduced environmental impact.

The rest of the paper is organized as follows. Section 2 summarizes related work in the literature. We introduce our system design in Sect. 3. The first prototype version is presented in Sect. 4. System evaluation is shown in Sect. 5. Finally, we conclude our paper in Sect. 6.

2   Related Work

Previous research in the field of environmental monitoring has explored various methods and technologies to collect data from aquatic environments. Manned boats equipped with specialized sensors were commonly used, but these approaches suffer from high operational costs, potential human risks, and limited adaptability to inaccessible or hazardous regions [7,17]. The need for skilled personnel to operate and manage these manned boats adds to the overall expenses and can pose safety concerns, especially in harsh weather conditions or when monitoring hazardous sites [8]. Additionally, the reliance on manned boats restricts the spatial coverage of data collection, leading to limitations in comprehensively assessing the environmental parameters across vast aquatic regions [9].

Remote sensing techniques, such as satellite-based data collection, have proven valuable for large-scale assessments but often lack the necessary resolution for fine-grained monitoring [6,18]. While satellite imagery can provide valuable insights into broad-scale environmental trends, it may not offer the required spatial and temporal resolution to capture fine-grained variations in water quality, especially in coastal and nearshore areas. Furthermore, the presence of cloud cover and atmospheric interference can limit the effectiveness of satellite-based monitoring, making it less reliable for real-time applications [7].

Recent advancements in autonomous vehicles, including USVs, have shown great promise in environmental research [3]. These autonomous vehicles offer several advantages over traditional manned approaches, including reduced operational costs and the elimination of human risks associated with field surveys (Johnson et al., 2021). Furthermore, USVs can access remote and hazardous areas that may be challenging or unsafe for human-operated vessels, expanding the possibilities for environmental data collection [13, 16].

Several studies have proposed the use of USVs for water quality monitoring, aquatic life surveying, and pollutant tracking [4, 15]. USVs equipped with a variety of sensors, such as turbidity meters, dissolved oxygen sensors, and chlorophyll fluorometers, have been successfully employed to monitor water quality parameters in lakes, rivers, and coastal regions [2]. Additionally, USVs have been deployed to survey and track aquatic organisms, including fish populations and marine mammals, providing valuable ecological insights [10].

However, the incorporation of IoT technology into these platforms to enable real-time data transmission, remote control, and sensor data fusion is an area that demands further exploration [11, 12]. Integrating IoT capabilities into USVs can enhance their environmental monitoring potential by enabling seamless data transmission to onshore base stations. Real-time data communication allows researchers to monitor aquatic environments continuously and respond promptly to any changes or emerging environmental issues. Moreover, IoT-based USVs can be remotely controlled and reprogrammed, enabling dynamic mission adjustments and optimizing data collection efficiency.

In summary, while previous research has explored various data collection methods in aquatic environments, there is a growing recognition of the potential of autonomous vehicles, particularly USVs, for environmental monitoring tasks. The integration of IoT technology into USVs opens up new possibilities for real-time, data-driven decision-making and resource management strategies [1, 14]. By capitalizing on the benefits of autonomous navigation, multi-sensor data collection, and IoT communication, the proposed IoT-based USV system holds the promise of advancing environmental research and addressing pressing ecological challenges in aquatic ecosystems.

3   System Design

The primary objective of the system is to enable vehicle control, offering the option for autonomous operation, pilot input through a ground control station, or interaction with an optional companion computer on board the vehicle. It is even possible to load a fully autonomous mission onto the vehicle for execution. The system consists of two main components: the USV (Unmanned Surface Vehicle) and the Ground Control Station. These components can communicate with each other using the Communication Layer. Figure 1 depicts the overview of the system based on ArduPilot.

Figure 2 depicts the detailed layers of the proposed system. The system consists of five layers as follows.

– Ground Station module: The Ground Station module serves as an essential external software integrated with ArduPilot, designed to operate on a computer stationed on the ground. Its primary function involves establishing seamless communication with an unmanned aerial vehicle (UAV). By facilitating real-time data exchange with the UAV, it becomes a crucial tool for both monitoring and controlling the aircraft's operations.

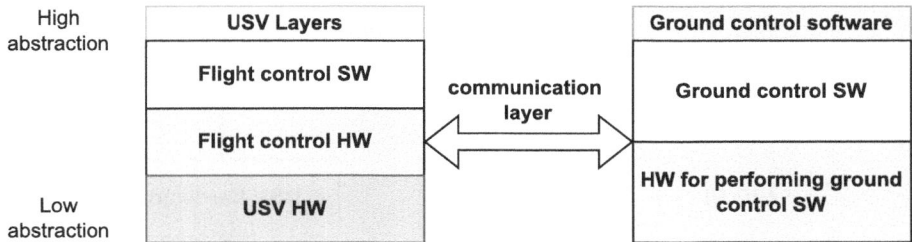

Fig. 1. Overview system for ArduPilot USV

– USV hardware: Another valuable addition to the suite of supported external software, DroneKit is specifically tailored to empower the creation of robust applications for UAVs. This software operates on the UAV's Companion Computer, unlocking the potential for advanced functionalities and enhanced capabilities within the vehicle's operational framework.
– MAVLink: An integral external software component deeply integrated into ArduPilot, MAVLink plays a vital role in ensuring effective communication between the system and various ground stations or companion computers. It boasts its own set of protocols that facilitate seamless and efficient data exchange, solidifying the foundation for a reliable and stable communication link.
– Vehicles modules: Central to the versatility and adaptability of ArduPilot, Vehicles Modules represent distinct firmware tailored for each specific vehicle type. This modular approach ensures that the system can be easily customized and optimized to suit the unique characteristics and requirements of diverse UAVs, catering to an extensive range of aerial platforms.
– Shared libraries modules: In an effort to promote code reusability and streamlined development, ArduPilot incorporates Shared Libraries Modules. These reusable code segments encompass core functionalities, sensor integration, and various libraries catering to a wide array of functions, such as control and navigation. By leveraging these shared modules, developers can efficiently build and expand upon the foundation of the system, fostering a collaborative and evolving ecosystem.
– Hardware abstractions layer: An indispensable feature of ArduPilot, the Hardware Abstractions Module ensures portability across an extensive selection of platforms and development boards. By providing a layer of abstraction between the underlying hardware and the software implementation, ArduPilot gains the flexibility to adapt to various setups, allowing developers to experiment and deploy the system on diverse hardware configurations with ease. This capability enhances accessibility and fosters innovation within the ArduPilot community, encouraging the exploration of novel applications and pushing the boundaries of UAV technology.

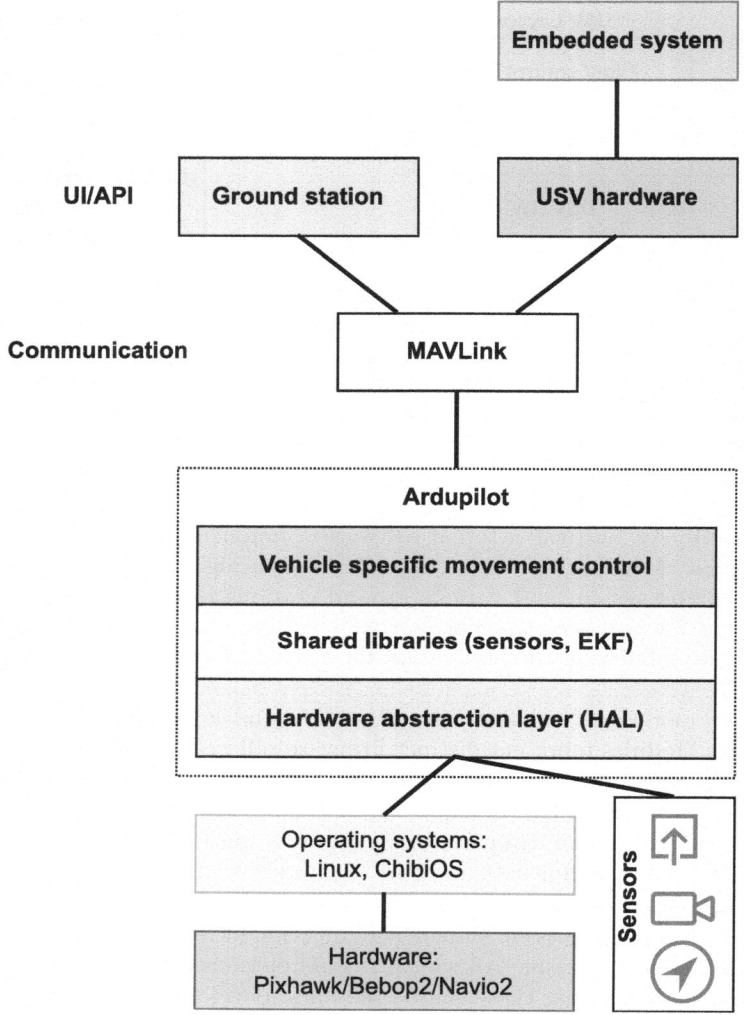

Fig. 2. Detailed layers of the proposed system

4   Prototype Implementation

The first prototype version of our USV is built with Raspberry pi 4 model B and the DroneKit-Python - an open source and community-driven project. For sensoring data, we use disolve oxygen, temperature, and PH level sensors.

DroneKit-Python is designed to be compatible with vehicles utilizing the MAVLink protocol for communication. With DroneKit-Python, developers gain the capability to build applications that operate on an onboard companion computer and establish seamless communication with the ArduPilot flight controller via a low-latency link. By leveraging on board apps, a wealth of opportunities arises to enhance the autopilot system by infusing greater intelligence into the vehicle's behavior. These apps can handle computationally intensive or time-sensitive tasks, such as computer vision, path planning, or 3D modeling, elevating the vehicle's capabilities to new heights.

Furthermore, DroneKit-Python is not limited to onboard applications alone; it also serves ground station applications. In this capacity, it enables communication with vehicles through a higher latency RF-link. This dual capability ensures a comprehensive and versatile toolkit for developers to innovate and create tailored solutions that suit both onboard and ground station requirements, ultimately advancing the realm of UAV technology. Figure 3 presents the GUI of the Dronekit project.

Fig. 3. The GUI of the Dronekit project [5]

5   Experimental Results

To evaluate the efficacy of the proposed system and its initial prototype version, we conducted field tests by deploying the Unmanned Surface Vehicle (USV) in the captivating Tram Chim National Park, nestled within the serene Mekong Delta region. Embracing the challenges of the forest environment, we successfully controlled the USV from a remarkable distance of 573 m using RF signal technology.

The evaluation further entailed examining the sensing capabilities of our USV, which produced exemplary results. The measured values demonstrated a strikingly low mean square error, indicating exceptional accuracy. When compared to handheld devices, our USV's sensor measurements exhibited a mere 6.2% variance, attesting to the robustness and reliability of our technology in diverse environmental conditions.

These encouraging findings validate the potential of our USV system, providing a strong foundation for its continued development and eventual application in a myriad of real-world scenarios, from ecological research in national parks to vital surveillance and monitoring tasks in various industries.

6   Conclusion

Water quality stands as a prominent environmental concern and ranks among humanity's most pressing global issues. The Mekong Delta, abundant in water resources from lakes, rivers, and aquifers, faces significant challenges due to dwindling usable water supplies. Water pollution and salinization have emerged as critical problems worldwide, attributable to factors such as oil spills, plastic waste, rising sea levels, and human activities. Such contamination poses a threat to aquatic habitats, fisheries, agriculture, and ultimately, human well-being.

This paper presents an innovative solution through the introduction of our IoT-based Unmanned Surface Vehicle (USV) designed specifically for monitoring the delicate wetland environment of the Mekong Delta. Leveraging recent advancements in open-source Global Positioning System (GPS)-guided drone technology, we have developed a cost-effective and easily transportable small unmanned surface vehicle (sUSV). Powered by Ardupilot open-source software, this agile USV is particularly suited for local scientists and marine managers, enabling them to comprehensively map and monitor marine environments in shallow areas with optimal visibility.

Equipped with an array of sophisticated sensors, the USV facilitates the measurement of various water parameters at multiple locations. Our experimental results showcase the impressive performance of the prototype version of the USV, capable of operating at distances as far as 573 m from the base station. Notably, the telemetry data from the sensors exhibit a mere 6.2% mean square error (MSE) when compared to readings obtained from certified handheld devices, reaffirming the accuracy and reliability of our technology.

By deploying our IoT-based Unmanned Surface Vehicle in the Mekong Delta, we strive to contribute significantly to the ongoing efforts of water quality management, ecological preservation, and sustainable resource utilization. This innovation represents a pivotal step forward in safeguarding vital ecosystems and promoting the well-being of both nature and humanity.

Acknowledgment. We acknowledge Ho Chi Minh City University of Technology (HCMUT), VNU- HCM for supporting this study.

References

1. Bălănescu, M., et al.: Study on unmanned surface vehicles used for environmental monitoring in fragile ecosystems. In: 2020 IEEE 26th International Symposium for Design and Technology in Electronic Packaging (SIITME), pp. 94–97. IEEE (2020)
2. Bogue, R.: The role of robots in environmental monitoring. Indust. Robot: Int. J. Robot. Res. Appl. **50**(3), 369–375 (2023)
3. Bovcon, B., Kristan, M.: A water-obstacle separation and refinement network for unmanned surface vehicles. In: 2020 IEEE International Conference on Robotics and Automation (ICRA), pp. 9470–9476. IEEE (2020)
4. Demetillo, A.T., Taboada, E.B.: Real-time water quality monitoring for small aquatic area using unmanned surface vehicle. Eng. Technol. Appl. Sci. Res. **9**(2) (2019)
5. Drone Kit: Welcome to DroneKit-Python's documentation. https://dronekit-python.readthedocs.io/en/latest/
6. Gall, M.P., Pinkerton, M.H., Steinmetz, T., Wood, S.: Satellite remote sensing of coastal water quality in Newzealand. NZ J. Mar. Freshwat. Res. **56**(3), 585–616 (2022)
7. Jorge, V.A., et al.: A survey on unmanned surface vehicles for disaster robotics: Main challenges and directions. Sensors **19**(3), 702 (2019)
8. Karthick, S., Kermanshachi, S., Namian, M.: Physical, mental, and emotional health of construction field labors working in extreme weather conditions: challenges and overcoming strategies. In: Construction Research Congress 2022, pp. 726–736 (2022)
9. Lazorchak, J.M., Klemm, D.J., Peck, D.V.: Environmental monitoring and assessment program surface waters: field operations and methods for measuring the ecological condition of wadeable streams (1998)
10. Levine, R.M., et al.: Autonomous vehicle surveys indicate that flow reversals retain juvenile fishes in a highly advective high-latitude ecosystem. Limnol. Oceanogr. **66**(4), 1139–1154 (2021)
11. Liu, R.W., et al.: Intelligent edge-enabled efficient multi-source data fusion for autonomous surface vehicles in maritime internet of things. IEEE Trans. Green Commun. Network. **6**(3), 1574–1587 (2022)
12. Prapti, D.R., Mohamed Shariff, A.R., Che Man, H., Ramli, N.M., Perumal, T., Shariff, M.: Internet of things (Iot)-based aquaculture: an overview of Iot application on water quality monitoring. Rev. Aquac. **14**(2), 979–992 (2022)
13. Sanfilippo, F., Tang, M., Steyaert, S.: The Aquatic Surface Robot (AnSweR), a lightweight, low cost, multipurpose unmanned research vessel. In: Yildirim Yayilgan, S., Bajwa, I.S., Sanfilippo, F. (eds.) Intelligent Technologies and Applications: Third International Conference, INTAP 2020, Gjøvik, Norway, September 28–30, 2020, Revised Selected Papers, pp. 251–265. Springer International Publishing, Cham (2021). https://doi.org/10.1007/978-3-030-71711-7_21
14. Sornek, K., et al.: Development of a solar-powered small autonomous surface vehicle for environmental measurements. Energy Convers. Manage. **267**, 115953 (2022)
15. Verfuss, U.K., et al.: A review of unmanned vehicles for the detection and monitoring of marine fauna. Mar. Pollut. Bull. **140**, 17–29 (2019)
16. Wang, W., Zhang, H., Li, Y., Zhang, Z., Luo, X., Xie, S.: Usvs-sim: a general simulation platform for unmanned surface vessels autonomous learning. Concurr. Comput.: Pract. Exp. **34**(3), e6567 (2022)

17. Yan, R.j., Pang, S., Sun, H.b., Pang, Y.j. J.: Development and missions of unmanned surface vehicle: Mar. Sci. Appl. **9**, 451–457 (2010)
18. Zhang, Z., Zhang, L., Wang, Y., Feng, P., He, R.: Shiprsimagenet: a large-scale fine-grained dataset for ship detection in high-resolution optical remote sensing images. IEEE J. Select. Topics Appl. Earth Observ. Remote Sens. **14**, 8458–8472 (2021)

3D CNN with BERT and Vision Transformer for Video Recognition

Bao Thai Duong[1] and Thai Hoang Le[2,3(✉)]

[1] Faculty of Information Technology, Ho Chi Minh City Open University, Ho Chi Minh City, Vietnam
bao.dt@ou.edu.vn
[2] Faculty of Information Technology, University of Science, Ho Chi Minh City, Vietnam
lhthai@fit.hcmus.edu.vn
[3] Vietnam National University, Ho Chi Minh City, Vietnam

Abstract. According to the development of the monitor system, detection and recognition are the major areas of interest within the field of computer vision. In recent years, due to their capacity to filter spatiotemporal video features, 3D CNN architectures with BERT have proven to be the best solution to this problem. Vision Transformer (ViT) has performed exceptionally well in recent benchmarks for image classification, object detection, and semantic image segmentation, among other computer vision applications. Transferring knowledge from such powerful ViT is an intriguing opportunity for developing excellent video recognition models. In this work, we discuss and evaluate the methods on HDMB-51 dataset to address the advantages and disadvantages. As a result, the study shows that two methods improve performance and accuracy of video recognition.

Keywords: Action Recognition · Video Recognition · Vision Transformers · 3D Convolution Neural Networks (3D CNNs)

1 Introduction

Action recognition is one of the most fundamental yet challenging tasks in video understanding. Human action recognition encompasses numerous computer vision research topics, such as medical supervision [17], micro video recommendation [18], autonomous driving [19], and so on.

Convolutional-based models that are optimized on the ImageNet dataset in a supervised fashion dominated this discipline over the past decade [1]. Based upon convolutional neural networks and now transformers, video recognition has achieved remarkable progress [20]. The CNN-based technique to extract meaningful characteristics from images is built around the convolution operation. Convolution operations include one-dimensional (1D), two-dimensional (2D), and three-dimensional (3D) convolution. Because of the importance of action recognition and other computer vision tasks in general, 2D and 3D convolution are naturally used more than 1D convolution for feature

© ICST Institute for Computer Sciences, Social Informatics and Telecommunications Engineering 2024
Published by Springer Nature Switzerland AG 2024. All Rights Reserved
P. Cong Vinh and N. Thanh Tung (Eds.): ICCASA 2023, LNICST 579, pp. 149–160, 2024.
https://doi.org/10.1007/978-3-031-58878-5_11

extraction. However, in action identification, where the goal is to acquire context and motion throughout the video, 3D convolution outperforms 2D convolution due to the capacity to collect spatiotemporal information in video at the same time.

Recently, researchers have shown an increased interest in Vision Transformer (ViT). ViT has a remarkable performance, good robustness, and smooth operation, which has received considerable critical attention in various visual recognition tasks such as image captioning, visual question answering, and multimodal understanding. CLIP, for example, can embed images and words into the same semantic space for similarity computation using 400M image-text pairs for training [5]. Furthermore, CLIP4Clip applies CLIP's image text expertise to the Video-Text Retrieval (VTR) problem, resulting in considerable performance increases across a variety of video-text retrieval datasets [6].

The objective of this study is to investigate the performance of 3D CNNs and ViT in the context of action recognition on video datasets. Both 3D CNNs and ViT have shown promising results in various computer vision tasks, but their effectiveness in action recognition remains an open question. In fact, the Human-Daily Activities with Multiple Cameras (HDMB-51) dataset is thought to be a good and famous dataset for action recognition tasks in the field of computer vision. It was designed to facilitate the study and evaluation of action identification algorithms and models. By conducting this comparison, we aim to shed light on the strengths and limitations of each method, ultimately providing insights into their suitability for action recognition tasks.

The rest of the paper is organized as follows. Section 2 discusses an overview of the related works in Sect. 2. A description of our approach is presented in Sect. 3. Section 4 details the experimentation carried out on. Finally, conclusions are given in Sect. 5.

2 Related Works

Over the past decade, convolutional networks have long been the standard architecture in video recognition. Supervised convolutional models that are optimized on the ImageNet dataset have dominated this discipline [1]. Simonyan et al. introduced a two-stream strategy in 2014 with the idea of having a CNN trained with raw RGB frames and another CNN trained with optical flow, which represents the movement vectors between two successive frames [10]. A combination of convolutional neural networks (CNNs) and long short-term memories (LSTMs) is presented by Wang et al. on untrimmed movies for poorly supervised action recognition and detection [14]. Convolutional neural networks (CNNs) have provided a high level of accuracy in image classification, so Krizhevsky et al. primarily employ this learning method among the other deep learning methods [4]. In 2018, Tran et al. discovered that disentangling spatial and temporal convolution improves the speed-accuracy tradeoff over the original 3D convolution [7]. Combining the 3DCNN and two-stream approaches is something that Wan et al. proposed by applying 3D convolutions to the spatial-stream and the VGG16 CNN to the temporal-stream. For the final prediction, they combined features from both streams and then utilized a support vector machine to improve accuracy to 70.2% [11].

Recently, Vision Transformers has emerged as a new trend in image recognition backbones. Transformers have also been adopted for video recognition. The Natural Language Processing scaling successes of Vaswani et al. [3]. Alexey Dosovitskiy et al.

presented a method for image recognition tasks based on transformers, which were originally devised for natural language processing (NLP) tasks [9]. Touvron et al. suggested a method for performing state-of-the-art picture classification tasks using fewer computer resources. They used knowledge distillation and provided data-efficient training approaches to boost the performance of smaller models [13]. The Swin Transformer, developed by Liu et al., employs a hierarchical architecture that grows quickly to accommodate high-resolution images. It employs shifting windows to properly capture local and global context [12]. Yuan-Hong Liao et al. investigated training Vision Transformers from scratch without using huge datasets like ImageNet for pretraining. Their strategy outperforms models pretrained on huge datasets in terms of performance [15]. Li et al. proposed the Vision Permutator, an architecture that employs permutation operations to replace self-attention layers in the traditional ViT model. It achieves competitive results with reduced computation [16].

Considering the above explorations, we propose BERT-based Temporal Modeling with 3D CNNs for action recognition tasks. By combining the advantages of 3D CNNs and BERT-based language modeling, this method efficiently captures the temporal dynamics and spatial aspects of films [21]. Additionally, several recent studies investigate ViT, which have drawn substantial attention and have displayed astounding performance in several computer vision applications [12, 13, 15, 16]. These methods have achieved the best possible results on a variety of standard datasets [22]. In Sect. 3 of the paper, we propose and describe two methods for evaluating the performance and accuracy of video recognition on the HDMB-51 dataset.

3   Methodology

3.1   Advantages BERT-Based Temporal Modeling with 3D CNNs

BERT-based Temporal Modeling with 3D CNNs provides several benefits for capturing temporal information and fusing textual and visual modalities for video comprehension. Combining the potential of BERT-based language modeling with 3D CNNs enables the capturing of temporal dependencies. This enables the model to comprehend the temporal relationships between frames or segments in a video, resulting in a more thorough analysis of the video's content. Second, the method enhances representations by combining the semantic understanding of BERT with the spatial-temporal features of 3D CNNs. The combination of these modalities improves the model's ability to capture the video's content and temporal context, resulting in representations that are richer and more informative. Moreover, BERT-based temporal modeling with 3D CNNs improves video comprehension by combining the strengths of textual and visual data. By combining these modalities, the model obtains a more comprehensive and holistic understanding of the video content, thereby enhancing its performance on tasks such as action recognition, video captioning, and video summarization. The approach also benefits from transfer learning, as the pretrained BERT model can capture high-level semantic information and the 3D CNNs can acquire specific spatial and temporal characteristics from video data. This transfer learning allows the model to apply knowledge from large-scale textual data to video tasks, despite having limited labeled video data. Overall, BERT-based Temporal

Modeling with 3D CNNs offers advantages in capturing temporal dependencies, enhancing representations, enhancing video comprehension, and facilitating transfer learning, making it a valuable technique for a variety of video-related applications.

3.2 Architecture of BERT-Based Temporal Modeling with 3D CNNs

The architecture of BERT-based Temporal Modeling with 3D CNNs is summarized in Fig. 1. It contains 3D CNN without applying temporal global average pooling. More details are provided in the original paper [8].

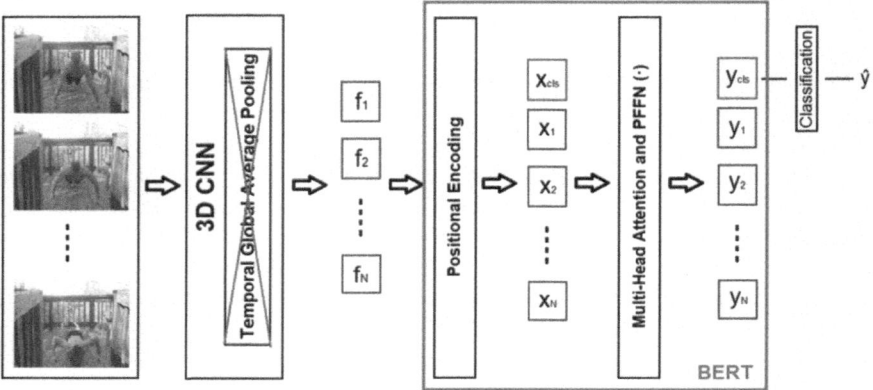

Fig. 1. The architecture of BERT-based Temporal Modeling with 3D CNNs [8].

3.3 Advantages Vision Transformer Model

Vision Transformer (ViT) has emerged as a formidable architecture in computer vision, offering several advantages over conventional convolutional neural networks (CNNs). The attention mechanism's capacity to capture global contextual information and long-range dependencies is a significant advantage. By considering the relationships between all image regions, the ViT is able to comprehend complex visual patterns and facilitate a holistic understanding of the visual content. In addition, ViT are scalable to large image resolutions, allowing for the efficient processing of high-resolution images without imposing substantial computational burdens. In addition, they offer versatility in handling variable image sizes, allowing for greater generalization and enhanced performance on datasets with varying aspect ratios. ViT's hierarchical representation learning, which enables the extraction of both local and global features by paying attention to various levels of abstraction, is an additional advantageous feature. ViT excels in transfer learning as well, as pretrained models can be fine-tuned for specific tasks, resulting in impressive performance across a broad range of computer vision applications. ViT provides interpretability and explain ability through their attention mechanism, allowing for insights into the model's decision-making process and fostering confidence in their predictions. Together, these benefits make ViT a compelling option for a variety of computer vision tasks and contribute to their rising prominence in the scientific community.

3.4 Architecture of Vision Transformer

Dosovitskiy et al. proposed the vision transformer (ViT), the first pure transform-architecture for image processing. It can achieve comparable outcomes to contemporary convolutional neural networks [9]. Figure 2 depicts the structure of ViT. There is a summary of the model: 1) divide an image into segments of fixed size. 2) embed each of them linearly 3) add position embeddings 4) provide the resultant vector sequence to a standard Transformer encoder. More information can be found in the original paper [9] (Fig. 2).

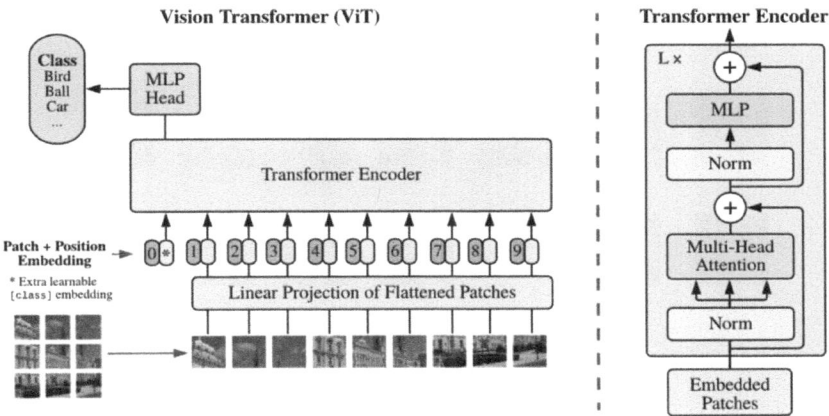

Fig. 2. Architecture of Vision Transformer [9]

3.5 Our Contribution

HDMB-51 dataset, which stands for "Human-Daily Activities with Multiple Cameras", is often considered suitable for action recognition tasks. Firstly, The HDMB-51 dataset comprises videos of fifty-one human daily activities captured from multiple camera viewpoints. It consists of many video clips, providing a substantial amount of training data. HDMB-51 focuses on fine-grained action recognition, requiring models to distinguish between subtle differences in human activities. The dataset is suitable to evaluate two methods as they provide the original videos. The HMDB-51 dataset is well-suited for evaluating 3D CNNs in action recognition tasks due to its video-based nature, temporal dynamics, diversity, and established role as a benchmark in the field. By using this dataset, researchers can effectively assess the capabilities of 3D CNNs in understanding complex actions in videos and drive advancements in the field of action recognition.

Moreover, ViT have shown promise in capturing fine-grained details, and their attention mechanisms allow them to address specific regions or frames in the video. ViT can effectively learn and generalize from this dataset, capturing both spatial and temporal information to recognize human activities with high accuracy. So, the HMDB-51 dataset is indeed suitable for evaluating 3D CNNs and ViT in action recognition tasks. Figure 3 shows the action categories in HDMB-51.

To demonstrate the effectiveness of the ViT and 3D CNNs, we propose two methods to evaluate based on the HDMB-51 dataset.

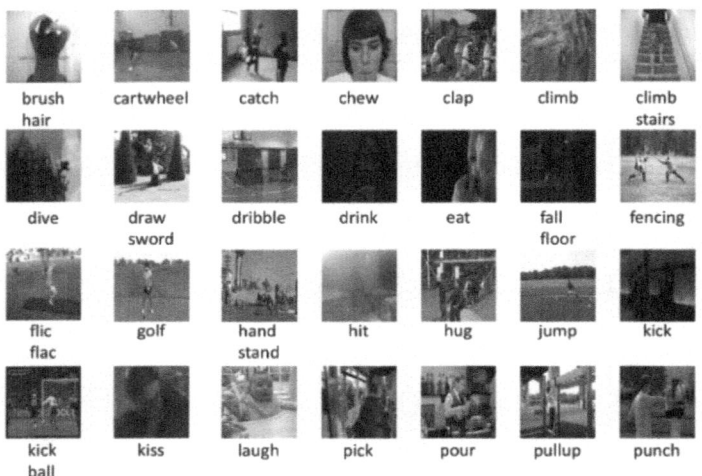

Fig. 3. Action categories in HDMB-51 [2]

We apply the general steps to apply ViT in video recognition tasks to demonstrate effective of ViT:

1. Data Preprocessing: Obtain the video dataset required for the task of recognition; Preprocessing the video entails removing frames from the video.
2. Temporal Tokenization: Extract or generate video clips with temporal context, ensuring they contain consecutive frames for each video.
3. Model Architecture: Select an appropriate ViT architecture for video recognition.
4. Model Pretrained: Initialize the ViT model with appropriate weights.
5. Model Evaluation: After training, evaluate the performance of the ViT on a validation or test set using relevant evaluation metrics.

The general steps to use ViT in video recognition are shown in Fig. 4.

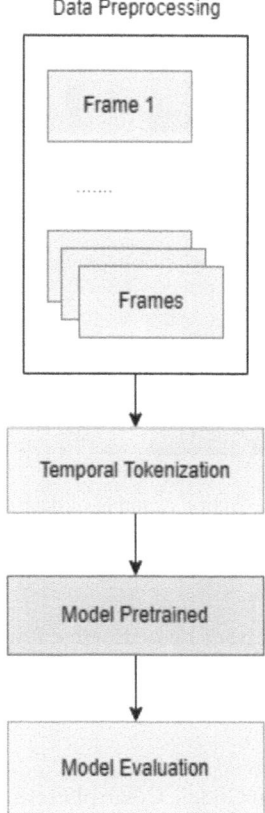

Data Preprocessing

Fig. 4. General step to apply ViT in video recognition.

4 Experiments

In this part, we talk about what happened when the experiment was done two different ways. First, we talk about the information, and then we get into the details of how it will be used. Then, we do studies to find out more about the two methods and compare them.

4.1 Dataset

In this section, we perform experiments on various settings, including zero-shot followed by ablation investigations of the proposed method. HMDB-51 is a compilation of realistic videos from various sources, including films and online recordings. The dataset contains approximately seven thousand video snippets organized into 51 action class categories [2]. HDMB-51 defines three data segments that are used to calculate the results. We report the mean accuracy of the three splits as the final accuracy (Fig. 5).

Fig. 5. Action example of dataset [2].

4.2 Dataset Preparation

The classification accuracy must be evaluated so that future classification outcomes may be predicted and compared. The operation flow is presented in Fig. 6.

4.3 Implementation

The paper runs experimentally on Google Colab platform with graphics processor 16 GB GPU P100, 12 GB RAM. Parameters are initialized in the same with the BERT-based Temporal Modeling with 3D CNNs 's original paper [8] and ViT 's original paper [9]. To save time and computational resources required for training, we use pre-trained models available, and it is compatible for action recognition [8, 9].

4.4 Result

We are able to produce the prediction score by combining the classification and selection scores. Table 1 and Fig. 7 present the five most similar labels for brush hair action.

4.5 Comparison with 3D CNN with BERT and ViT

From the values on Table 2, the result shows that the two methods show their robustness by performing efficiently in terms of action recognition. It can be shown that the ViT (ViT-B/16) produced particularly good results by achieving 84,7% accuracy. ViT-B/16 outperformed ResNeXt101 BERT slightly, indicating its potential as a strong contender in the domain of action recognition.

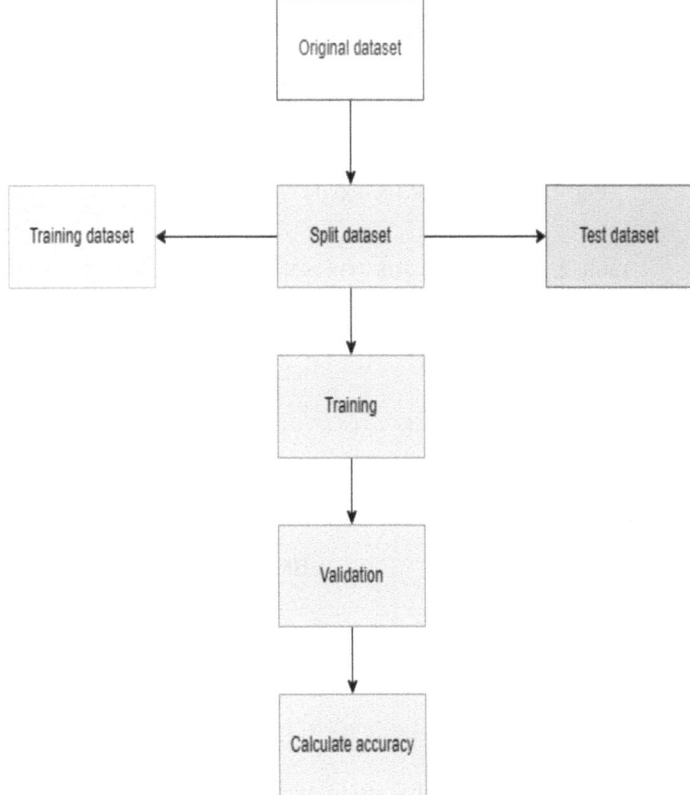

Fig. 6. The dataset's structure and operation flow

Table 1. Evaluation by using ViT

brush hair	77,2%
clap	18,3%
cartwheel	3,2%
dribble	0,6%
catch	0,1%

4.6 Zero-Shot Experiments

Zero-shot learning is a machine learning paradigm in which a model is trained to recognize and generalize to classes that were not seen during the training phase. We use ResNeXt101 BERT and ViT-B/16 to perform cross-dataset zero-shot evaluation in video dataset. We present a comprehensive comparison in Table 3.

Top 1: brush hair
Top 2: clap
Top 3: cartwheel
Top 4: dribble
Top 5: catch

Fig. 7. The result prediction on video by using ViT.

Table 2. Comparison with 3D CNN with BERT and ViT

Method	HMDB-51
ResNeXt101 BERT	83.5%
ViT-B/16	84.7%

Table 3. Zero-shot performance on HDMB-51

Method	HMDB-51
ResNeXt101 BERT	40.8%
ViT-B/16	44.6%

4.7 Discussion

Our study shows that ViT is a powerful video recognition architecture. ViT scale and adapt to diverse video resolutions better than 3D CNNs. Despite both architectures being pretrained on the Kinetics-400 dataset [23], we observed that the ViT outperformed the 3D CNN in terms of performance on the task of action recognition. While 3D CNNs are still effective, ViT offer a promising alternative, and their performance on the HDMB-51 dataset shows their versatility and potential in video understanding.

5 Conclusion

In this paper, we propose to use 3D CNNs and ViT which aim to verify the effectiveness of the methods. 3D CNNs were traditionally the dominating architecture for computer vision applications like image classification, object recognition, and segmentation. ViT produced impressive results, even outperforming 3D CNNs in some circumstances, particularly when trained on large-scale datasets. Moreover, ViT can process films of various sizes without increasing processing overhead by tokenizing video frames into patches. So, ViT can handle films of varying resolutions and aspect ratios, making them useful for video recognition jobs. Besides, ViT has some challenges with computational complexity and spatial information loss. In the future, we intend to extend our method beyond categorization to other video tasks. In the future, we will develop a system to detect the features of the actions.

References

1. Deng, J., Dong, W., Socher, R., Li, L.-J., Li, K., FeiFei, L.: Imagenet: a large-scale hierarchical image database. In: CVPR, pp. 248–255 (2009)
2. Kuehne, H., Jhuang, H., Garrote, E., Poggio, T., Serre, T.: HMDB: a Large Video Database for Human Motion Recognition. In: ICCV (2011)
3. Vaswani, A., et al.: Attention is all you need. In: NeurIPS, pp. 5998–6008 (2017)
4. Krizhevsky, A., Sutskever, I., Hinton, G.E.: ImageNet classification with deep convolutional neural networks. Commun. ACM **60**(6), 84–90 (2017). https://doi.org/10.1145/3065386
5. Radford, A., et al.: Learning transferable visual models from natural language supervision. Image 2, T2 (2021)
6. Luo, H.S., et al.: CLIP4Clip: An empirical study of clip for end-to-end video clip retrieval and captioning. Neurocomputing **508**, 293–304 (2022). https://doi.org/10.1016/j.neucom.2022.07.028
7. Tran, D., Wang, H., Torresani, L., Ray, J., LeCun, Y., Paluri, M.: A closer look at spatiotemporal convolutions for action recognition. In: CVPR (2018)
8. Esat Kalfaoglu, M., Sinan Kalkan, A., Alatan, A.: Late temporal modeling in 3D CNN architectures with bert for action recognition. In: Bartoli, A., Fusiello, A. (eds.) Computer Vision – ECCV 2020 Workshops: Glasgow, UK, August 23–28, 2020, Proceedings, Part V, pp. 731–747. Springer International Publishing, Cham (2020). https://doi.org/10.1007/978-3-030-68238-5_48
9. Dosovitskiy, A., et al.: An image is worth 16×16 words: Transformers for image recognition at scale. In: Proceedings of the 9th International Conference on Learning Representations (2021)
10. Simonyan, K., Zisserman, A.: Two-Stream Convolutional Networks for Action Recognition in Videos. Adv. Neural. Inf. Process. Syst. **27**, 1–9 (2014)
11. Wan, Y., Yu, Z., Wang, Y., Li, X.: Action recognition based on two-stream convolutional networks with long-short-term spatiotemporal features. IEEE Access **8**, 85284–85293 (2020). https://doi.org/10.1109/access.2020.2993227
12. Liu, Z., et al.: Swin transformer: hierarchical vision transformer using shifted windows. In: Proceedings of the IEEE/CVF International Conference on Computer Vision (ICCV), pp. 10012–10022 (2021)
13. Touvron, H., Cord, M., Douze, M., Massa, F., Sablayrolles, A., Jegou, H.: Training data-efficient image transformers & distillation through attention. In: Proceedings of the 38th International Conference on Machine Learning, PMLR vol. 139, pp. 10347–10357 (2021)
14. Wang, L., Xiong, Y., Lin, D., Van Gool, L.: Untrimmednets for weakly supervised action recognition and detection. In: CVPR (2017)
15. Yuan, L., et al.: Tokens-to-token ViT: Training vision transformers from scratch on imagenet (2021). arXiv:2101.11986. Retrieved from https://arxiv.org/abs/2101.11986
16. Hou, Q., Jiang, Z., Yuan, L., Cheng, M.M., Yan, S., Feng, J.: Vision Permutator: A Permutable MLP-Like Architecture for Visual Recognition. IEEE Trans. Pattern Anal. Mach. Intell. **45**(1), 1328–1334 (2023). https://doi.org/10.1109/TPAMI.2022.3145427
17. Hershberger, W.A.: Chapter 1 The Synergy of Voluntary and Involuntary Action. In: Volitional Action - Conation and Control, vol. 62, pp. 3–20. Elsevier (1989). https://doi.org/10.1016/S0166-4115(08)61905-6
18. Zhu, Y., et al.: A comprehensive study of deep video action recognition (2020). arXiv preprint arXiv:2012.06567
19. Herath, S., Harandi, M., Porikli, F.: Going deeper into action recognition: A survey. Image Vis. Comput. **60**, 4–21 (2017). https://doi.org/10.1016/j.imavis.2017.01.010

20. Karpathy, A., Toderici, G., Shetty, S., Leung, T., Sukthankar, R., Fei-Fei, L.: Large-scale video classification with convolutional neural networks. In: CVPR, pp. 1725–1732 (2014)
21. Devlin, J.: Bert: Pre-training of deep bidirectional transformers for language understanding (2018). arXiv preprint arXiv:1810.04805
22. Tummala, S., Kadry, S., Bukhari, S.A.C., Rauf, H.T.: Classification of brain tumor from magnetic resonance imaging using vision transformers ensembling. Curr. Oncol. **29**(10), 7498–7511 (2022). https://doi.org/10.3390/curroncol29100590
23. Kay, W., et al.: The kinetics human action video dataset (2017). arXiv preprint arXiv:1705. 06950

Identify Tumors on Lung CT Images

Phong Thanh Le[1,2], Thai Hoang Le[1,2(✉)], and Hieu Duc Thai Tran[1,2]

[1] Faculty of Information Technology, University of Science, Ho Chi Minh City, Vietnam
lhthai@fit.hcmus.edu.vn
[2] Vietnam National University, Ho Chi Minh City, Vietnam

Abstract. This paper introduces You Only Look Once (YOLO) model to identify tumors on computed tomography (CT) lung images. The model uses a variant of the YOLO algorithm called YOLOv5 [1], which is known for its accuracy and speed in object detection tasks. To train and evaluate the YOLO model, we use the Lung Nodule Analysis 2016 dataset (LUNA16) [2]. This dataset contains a set of lung CT scans with annotations indicating the location of the tumors. We preprocess the CT images and annotations to prepare data for model training and testing. During the training phase, the YOLO model uses a loss function named Generalized Intersection over Union (GIoU) loss [3], which provides a more accurate measure of box overlap between the predicted objects and the ground truth. The combination of the YOLO architecture and the GIoU loss enables accurate and fast detection, making the proposed model a promising tool to aid physicians in diagnosing lung cancer.

Keywords: You Only Look Once · Tumor identification · Lung Computed Tomography (CT) images · Lung Nodule Analysis 2016 (LUNA16) dataset · Generalized Intersection over Union (GIoU) loss · Diagnosis Lung cancer

1 Introduction

Lung cancer remains a major global health concern as it continues to be the most common form of cancer and is associated with the highest number of cancer-related deaths worldwide [4]. Finding and diagnosing lung cancer at an early stage is critical to improving patient outcomes and improving the effectiveness of treatment options. To assist doctors in the diagnostic process, CAD (Computer Aided Detection) systems [5, 6] have been developed. These systems use advanced image processing algorithms and techniques to assist radiologists and other medical professionals in the analysis and interpretation of medical images, especially those Images are obtained through techniques such as Computed Tomography (CT). The CAD system for lung cancer typically consists of two main stages. The main stage is to detect candidate nodules, which are potential tumor regions within the lung. This phase involves applying complex algorithms to segment lung images and identify suspicious areas that may be indicative of nodules. Once candidate nodules are identified, the CAD system proceeds to classify them as

P. Cong Vinh and N. Thanh Tung (Eds.): ICCASA 2023, LNICST 579, pp. 161–174, 2024.
https://doi.org/10.1007/978-3-031-58878-5_12

positive, indicating the presence of a tumor. This classification stage is crucial in distinguishing between cancerous and noncancerous nodules. The CAD system not only detects and classifies nodules, but can also provide additional information about their characteristics. For example, it can help determine the size, location, shape irregularities, and other characteristics of identified nodules. This information is valuable in assisting physicians with treatment planning and decision making. Furthermore, CAD systems can assist in monitoring the growth or change of nodules over time, providing valuable insights into disease progression and treatment response. Machine learning techniques, especially deep learning algorithms, have revolutionized the analysis of medical images, including the detection and classification of lung nodules for the diagnosis of lung cancer. These algorithms are ideally suited for tasks involving pattern recognition in large datasets, making them highly effective in medical imaging applications. By leveraging the power of CAD systems and machine learning algorithms, medical professionals can improve the detection and diagnosis of lung cancer in its early stages, leading to more effective treatment strategies. And better outcomes for patients. As research and technology continue to advance, CAD systems are expected to play an increasingly important role in the fight against lung cancer and other medical conditions. In the field of tumor detection and identification on lung CT images, there are numerous methods available, including Computer-Aided Diagnosis (CAD) systems. These systems leverage advanced algorithms to assist radiologists and medical professionals in the accurate and efficient detection of tumors and abnormalities in medical images. For the specific task at hand, we have chosen to implement a new deep learning model as part of our CAD system. Deep learning has shown significant promise in various medical imaging tasks, including tumor detection, due to its ability to automatically learn and extract relevant features from the data. In the next section of our paper, we plan to provide a detailed description of our new deep learning model. The rest of the paper is organized as follows. Section 2 discusses an overview of the related works in Sect. 2. A description of our approach for is presented in Sect. 2. Section 3 details the experimentation carried out on. Finally, conclusions are given in Sect. 4

2   Related Work

2.1   Object-Detection [7] Algorithms

Deep learning models have demonstrated outstanding potential in detecting lung tumors, offering promising avenues for advancing medical imaging analysis. However, to understand their true performance in clinical settings, insights from medical professionals and researchers are indispensable. Medical professionals possess invaluable domain knowledge and practical experience in diagnosing lung tumors from CT images, making their perspectives important in evaluating the effectiveness of deep learning models. In this context, challenges in tumor identification arise in specific situations, such as distinguishing between benign and malignant tumors or detecting nodules in complex cases. Each method can excel under different conditions, requiring evaluation based on specific use cases and datasets. Swetha Subramanian [8] proposed a 50 x 50 pixel image-based tumor identification procedure, focusing on tumor identification using small arrays of images. This approach is capable of using a Convolutional Neural Network (CNN) architecture specifically designed for image analysis tasks. On the other hand, Ding et al. [9] proposed a 3-D CNN-based lung nodule detection system, using the Pulmonary Nodule Analysis Challenge (LUNA16) dataset, containing CT scans with annotations for lung nodules (potential tumors). It is important to note that these models target different aspects of tumor identification, with Swetha Subramanian's model focusing on smaller arrays of images, while Ding et al. using 3-D CNN to analyze the entire volumetric CT image. To further compare the performance of the object detection models, we consider the Mean Accuracy (mAP) (Fig. 1). Faster algorithms like Single Shot Multibox Detector (SSD) may struggle with the complexity of the tumor identification problem. On the other hand, Faster R-CNN [10] prioritizes accuracy over speed and can achieve six times better performance than SSDs with the same feature extractor. Even though YOLO v5 exhibits improved performance over SSDs and Fully Area-Based Coherent Network (RFCN), Faster R-CNN maintains a small advantage of 0.1 mAP at 4,000 training steps. Given the diverse strengths and challenges associated with different deep learning models for tumor recognition, the decision was made to use YOLO v5x as the official technique for testing and evaluation to demonstrate its effectiveness. Superior performance compared to other object detection models such as SSD and RFCN. By using YOLO v5x, we aim to take advantage of its ability to efficiently detect lung tumors from CT images. We recognize the importance of accurate tumor identification in the clinical setting, especially when distinguishing between benign and malignant tumors or dealing with complex cases. The perspective and field knowledge of medical professionals informed our decision to select a model that could excel at detecting tumors from small arrays of images, which could be particularly beneficial in identifying subtle abnormalities and improving early detection rates.

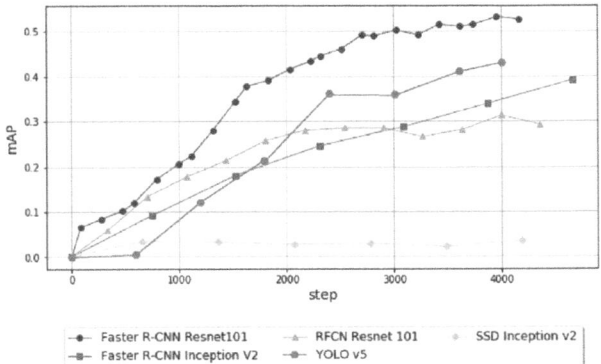

Fig. 1. Performance comparison using the mAP of the object-detection models

2.2 YOLO v5x

2.2.1 Architecture

YOLO v5x (You Only Look Once version 5) is an advanced object detection algorithm that builds on the success of its predecessors, most notably YOLOv3. It represents a significant improvement in both speed and accuracy. The key innovation of YOLO v5x lies in its single-stage approach, which allows it to predict bounding boxes and class probabilities for objects directly in a single neural network pass. This design not only reduces computational complexity, but also enables real-time object detection with impressive accuracy, making it well-suited to a wide range of applications, including mission critical is to detect lung nodules in medical imaging. The architecture of YOLO v5x consists of three essential components (Fig. 2), each serving a specific purpose.

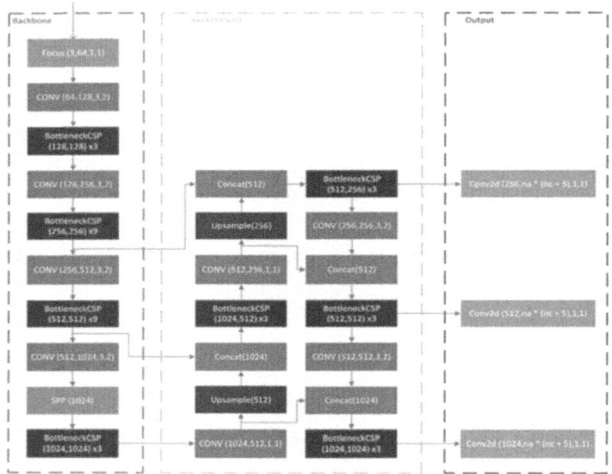

Fig. 2. Architecture of YOLO v5x

Backbone model is a fundamental part of YOLO v5x, responsible for extracting rich and informative features from input image. This process involves passing the image through several convolutional layers, capturing hierarchical features of increasing complexity. By learning relevant patterns and characteristics from images, Backbone models provide the foundation for accurate object detection. The Neck model is an intermediate component responsible for generating the characteristic pyramids. Feature pyramids are multi-scale representations of images that help identify objects of different sizes and scales in the input. This aspect is especially important in medical imaging, as lung nodules can vary considerably in size and shape. The Neck model's ability to generate pyramidal pyramids enhances YOLO v5x's ability to efficiently detect lung nodules of different sizes. The Head model is the final component of YOLO v5x and is responsible for making the final output predictions. It takes the features extracted by the Backbone model and further refines them based on the feature pyramids generated by the Neck model. The Head model then predicts bounding boxes consisting of detected objects, along with corresponding class probabilities for each feature type. In the context of lung nodule detection, the Head model's predictions pinpoint the location and type of nodule in the CT scan. By integrating information from the Backbone and Neck models, YOLO v5x can efficiently and accurately detect objects, including lung nodules, of various sizes and scales within an input CT image. The use of feature pyramids generated by the Neck model contributes to the algorithm's robust object identification capabilities. Moreover, YOLO v5x leverages the concept of anchor boxes, combined with class probabilities, in the Head model to further enhance the precision and reliability of the final predictions. These anchor boxes act as reference templates for the algorithm to detect and localize objects more accurately. The end-to-end design of YOLO v5x enables both speed and accuracy in tumor detection, making it an incredibly promising choice for medical image analysis tasks, particularly tumor identification in lung CT images. Its real-time capabilities facilitate rapid analysis of large volumes of medical data, enabling timely diagnosis and treatment planning, which is crucial for ensuring the best possible patient outcomes.

2.2.2 Advantages

YOLO v5x offers several advantages that make it a powerful and efficient object detection algorithm, particularly for applications like tumor identification in lung CT images. YOLO v5x is a remarkable advancement in object detection algorithms, building upon the success of its predecessors and introducing several improvements to enhance its overall performance. One of its key strengths is its ability to achieve higher accuracy and better detection results compared to previous versions, such as YOLOv3. This enhanced performance is particularly critical for accurate tumor identification in lung CT images and can significantly impact the effectiveness of lung cancer diagnosis. One of the standout features of YOLO v5x is its optimization for speed, enabling real-time object detection even on resource-constrained devices. Its lightweight architecture allows for fast inference without compromising accuracy, making it well-suited for applications that require quick responses. In the context of lung CT image analysis, real-time inference can expedite the tumor identification process, leading to faster diagnosis and more prompt treatment planning. Furthermore, YOLO v5x's high level of customizability empowers researchers and developers to fine-tune the model to suit specific use cases. They can modify the architecture, backbone, and other components to adapt the algorithm to different tasks and datasets. This flexibility is especially advantageous in medical imaging, where customizing the algorithm to the unique characteristics of lung CT images can lead to improved performance and better tumor identification. Another crucial advantage of YOLO v5x is its ability to handle images of varying resolutions and aspect ratios without requiring fixed input image sizes. This adaptability is particularly beneficial when dealing with medical images that may have different dimensions or varying pixel densities. The algorithm's capability to handle diverse image sizes enhances its applicability to lung CT images, which often come in various resolutions. In challenging and cluttered scenes, YOLO v5x excels in detecting objects, including tumors. It can effectively handle overlapping objects and accurately identify small or distant tumors, a crucial feature in medical imaging applications. This robustness significantly contributes to accurate tumor identification in diverse clinical scenarios. The utilization of anchor boxes and efficient regression techniques in YOLO v5x results in more accurate and precise bounding box predictions around detected objects. This level of object localization is paramount in medical applications, where accurate tumor boundaries are vital for accurate diagnosis and effective treatment planning. Precise localization empowers medical professionals to confidently assess tumor characteristics for making well-informed treatment decisions. YOLO v5x follows a single-stage detection approach, eliminating the need for proposal generation and subsequent refinement stages. This simplicity reduces computational overhead and enables faster inference. In the context of lung tumor identification, this streamlined process allows for rapid analysis of CT scans, making YOLO v5x an excellent choice for time-sensitive medical applications. The fact that YOLO v5x is open-source and comes with pre-trained models and extensive documentation makes it accessible and user-friendly for both researchers and developers. Its straightforward interface simplifies the training and deployment process, facilitating seamless integration into existing medical imaging workflows.

2.3 Our Contribution

To demonstrate the effectiveness of YOLO v5x in the domain of tumor identification, we propose a novel model that leverages the advanced capabilities of YOLO v5x in object detection. Our primary objective is to develop a system that can accurately and efficiently detect tumors on lung CT images, contributing to early tumor detection and improved patient outcomes in the context of lung cancer diagnosis. The development of our tumor identification model involved a series of well-defined steps, illustrated in Fig. 3.

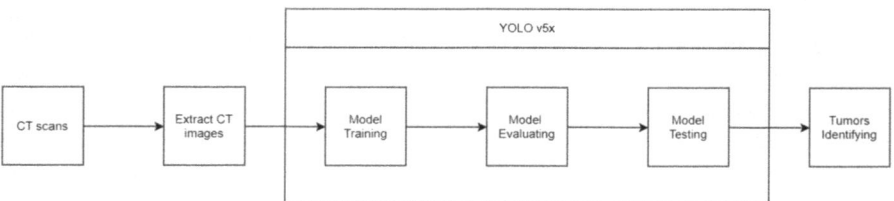

Fig. 3. Our model identifies tumors on lung CT images.

In Fig. 3, we began our tumor identification model by meticulously sorting and selecting an extensive dataset that included a diverse array of lung CT scans with annotated tumor regions. The quality, size, and diversity of the dataset are of prime importance in ensuring the robustness and generalizability of our model. To prepare the dataset for training, we used the necessary data preprocessing steps, including image resizing, normalization, and enhancement. These steps serve to normalize the input data, improve the performance of the model, and enhance the training data, thus increasing its diversity. Then, the custom YOLO v5x model underwent exhaustive training on the selected dataset, where we optimized its parameters to maximize the tumors identification accuracy. The training process involves iterative updates to the model based on the calculated loss, which is mainly driven by the GIoU loss function. This loss function ensures precise bounding box positioning, allowing the model to accurately and accurately delineate tumor regions on lung CT images. An important aspect of YOLO v5x is its optimized speed for real-time object detection. To showcase the real-time inference capabilities of our model, we demonstrated the ability to rapidly process and identify tumors on lung CT images. This property makes our model well-suited to time-sensitive medical situations where rapid tumors identification is critical for prompt patient care. Furthermore, to ensure the model's adaptability and generalization to real-world data, we incorporated a fine-tuning and adaptability process. This allows our model to handle variations in the data set, such as different lung nodule types, sizes, and image quality, thus improving the model's performance in different cases. Different .

When using our model (Fig. 3) for tumor identification on lung CT images with the LUNA16 dataset, several significant contributions can be observed in the context of medical imaging and lung cancer diagnosis. Our model, with its improved performance and architecture, can achieve high accuracy in identifying lung tumors within the LUNA16 dataset. The algorithm's ability to handle various image sizes and complex scenarios in CT scans leads to more precise tumor localization and reduced false positives and false negatives. Our model optimized speed allows for real-time tumor detection, making it well-suited for analyzing large volumes of CT scans from the LUNA16 dataset efficiently. This real-time capability ensures rapid screening and diagnosis, which is crucial in medical settings for timely patient care. By accurately identifying small and subtle tumors in lung CT images from LUNA16, our model contributes to early detection. Early identification of lung tumors can lead to improved treatment outcomes and potentially save lung. Our model serves as a valuable tool to assist radiologists and healthcare professionals in the tumor identification process. Its accurate predictions and automated detection can speed up the screening process and provide valuable insights, allowing medical experts to focus on critical decision-making and patient care. Our model can be fine-tuned and customized to suit the specific characteristics of the LUNA16 dataset, such as the nature of lung nodules and variations in imaging quality. This adaptability ensures better generalization to the LUNA16 data, leading to improved performance on this specific dataset. Our model precise tumor localization in lung CT scans from LUNA16 provides crucial information for treatment planning. Accurate tumor boundaries aid in developing targeted treatment strategies, optimizing therapy delivery, and minimizing damage to healthy tissues. Integrating our model into clinical workflows with the LUNA16 dataset can streamline tumor identification processes. The algorithm's user-friendly interface and real-time inference allow for efficient analysis and diagnosis, saving time and resources for medical professionals. Our model's application to the LUNA16 dataset contributes to the advancement of research in lung cancer detection and medical image analysis. The insights gained from this study can lead to further improvements in lung nodule detection algorithms and drive innovations in the field. Our model's efficient and accurate tumor detection capabilities make it a valuable asset for large-scale lung cancer screening programs. By processing a significant number of CT scans, the algorithm can support population-based cancer screening initiatives, leading to early tumor detection and improved public health outcomes. The results of our experiments demonstrate the real-world efficacy of our model in tumor detection. During the evaluation process, we utilized a carefully curated and diverse dataset, representative of the complexity and variety of lung nodules encountered in clinical practice.

3 Experiments Results

3.1 Datasets

Lung Nodule Analysis 2016 (LUNA16) is a significant dataset used in the field of lung nodule detection and analysis. It comprises 888 CT scans, which are three-dimensional medical imaging data obtained from computed tomography machines. These CT scans are crucial in identifying and diagnosing lung nodules, which are potential indicators of lung cancer. In the LUNA16 dataset, a file is provided that contains an extended set of candidate locations for the 'false positive reduction' track. False positives are instances where the algorithm incorrectly identifies a region as a nodule when it is not actually present. To address this, the dataset includes a set of candidate locations to help improve the accuracy of nodule detection algorithms by filtering out false positives. It is important to note that a single nodule can be detected by multiple candidates. This is because lung nodules can vary in size, shape, and appearance, leading to multiple possible detection points within a single nodule. In the LUNA16 dataset, a total of 1186 nodules have been detected, and the number of candidates generated for these nodules amounts to 754,975. This large number of candidates provides a diverse set of potential nodule locations for the algorithm to analyze and evaluate. Each CT scan in the LUNA16 dataset consists of a series of 200 to 400 individual images, known as slices. These slices are two-dimensional representations of the three-dimensional CT scan data. Each slice has a size of 512 x 512 pixels, which represents the image resolution and dimensions.

Based on the author [8], we used bounding boxes with a fixed size of 50 x 50 pixels to train, validate, and test our model for lung nodule detection. Bounding boxes are rectangular regions that enclose the regions of interest in an image, in this case, the lung nodules. The size of 50 x 50 pixels was chosen as a standard region to focus on the nodules and provide a consistent input for the model. The dataset used for training and evaluation consists of a total of 5486 lung nodules with positive labels. These positive labels indicate the presence of nodules in the corresponding images. The dataset is divided into three subsets: 4149 images for training, 977 images for validation, and 1418 images for testing. During the training phase, our model is presented with the 50 x 50 pixel gray-scale images of lung nodules along with their corresponding bounding box annotations. The model learns to detect and localize nodules within the images, effectively predicting the bounding boxes that enclose the nodules. The validation subset, comprising 977 images, is used during the training process to monitor the model's performance and prevent overfitting. This process helps ensure that the model generalizes well to new, unseen data, rather than memorizing specific examples from the training set. Finally, the testing subset, containing 1418 images, is employed to assess our model's performance on the data. The model's predictions on this test set are evaluated against ground-truth annotations to measure the algorithm's accuracy, precision, recall, F1-score, and other performance metrics. This step helps validate the model's effectiveness in detecting lung nodules in real-world scenarios. By using gray-scale images and bounding box annotations of fixed size, we achieve consistency in data representation, making the training process more manageable and efficient. The use of a large dataset with a significant number of nodules ensures that our model can learn from diverse examples, leading to robust and accurate nodule detection.

3.2   Environment for Experiments

We conducted the implementation and execution of our model in a consistent environment using the Google Colab platform. The platform provided a 16 GB GPU A100 and 12 GB RAM, which was crucial for efficient training and inference of the deep learning model. To initialize the model, we used a set of weights based on the YOLO v5x architecture. Weight initialization is essential as it helps the model start with reasonable parameter values, which can expedite the training process and lead to more stable convergence. During the training phase, we employed the Adam optimizer with a learning rate of 1e-2 and a weight decay of 5e-4. Adam is a popular optimization algorithm that efficiently adapts the learning rate for each parameter, making it well-suited for training deep neural networks like YOLO v5x. During the training phase, we employed the Adam optimizer with a learning rate of 1e-2 and a weight decay of 5e-4. Adam is a popular optimization algorithm that efficiently adapts the learning rate for each parameter, making it well-suited for training deep neural networks like YOLO v5x. To improve the model's generalization and robustness, we applied data augmentation techniques during training. We adjusted the hue, saturation, and value (brightness) of the images with hyperparameters set at 0.015 for hsv_h, 0.7 for hsv_s, and 0.4 for hsv_v. This technique introduces variations in color to make the model more resilient to changes in illumination and contrast. We performed random translations of the images with a hyperparameter set to 0.1. This technique shifts the objects' positions within the images, simulating different viewpoints and object locations. We applied random scaling to the images with a hyperparameter set to 0.5. This technique allows the model to detect objects at various sizes, which is essential for handling lung nodules of different dimensions. We used mosaic augmentation with a hyperparameter set to 1. Mosaic augmentation combines four random images into a single mosaic image, providing the model with more complex and diverse examples for training. We trained the model for 100 epochs, with a batch size of 16 images. The number of epochs determines the number of times the entire dataset is processed during training. Training for 100 epochs allows the model to learn from the data effectively and potentially achieve convergence. To prevent overfitting and avoid wasting training time once the model has converged, we employed the early stop technique. Early stop enables the model to stop training when its performance on the validation set starts deteriorating, indicating that further training may lead to overfitting. By implementing the YOLO v5x models in the specified environment and applying appropriate hyperparameters, data augmentation, and early stop technique, we aimed to achieve robust and accurate lung nodule detection in lung CT images using the LUNA16 dataset.

3.3   Evaluation

We developed a two-step approach to train and evaluate our model for lung nodule detection in lung CT images using the YOLO v5x architecture. In the first step, we used a pre-training model and the training data to find the initial parameters for our model. The pre-training model might have been a pre-trained YOLO v5x model on a large dataset, such as the COCO dataset, to capture general object detection capabilities. We fine-tuned this pre-training model on our specific dataset, which consists of lung CT images with labeled lung nodules. During training, we utilized the back-propagation method, a standard optimization technique for training neural networks. Back-propagation calculates gradients with respect to the model parameters, and the model iteratively adjusts these parameters to minimize the difference between predicted bounding boxes and ground-truth annotations of lung nodules. This process allows the model to learn the patterns and features relevant to lung nodule detection. After training, we used the cross-validation method to assess the model's accuracy and generalization performance. Cross-validation involves splitting the training data into multiple subsets, or "folds," and iteratively using each fold as a validation set while training the model on the remaining folds. This process is repeated to obtain more robust estimates of the model's performance. During cross-validation, we measured several performance metrics, including Precision, Recall, and F1-score. Precision represents the ratio of true positive detections (correctly identified nodules) to all positive detections (both true and false positives). Recall, also known as Sensitivity or True Positive Rate, measures the ratio of true positive detections to the total number of actual nodules in the dataset. The F1-score is the harmonic mean of Precision and Recall, providing a single metric that balances both metrics and gives an overall indication of the model's performance. In the final step, we evaluated the trained model's accuracy on the testing data, which the model has never seen during training or cross-validation. This evaluation allowed us to measure the model's performance on unseen data, providing insights into its ability to generalize to new lung CT images with lung nodules. By assessing Precision, Recall, and F1-score during cross-validation and testing, we obtained a comprehensive evaluation of our model's performance for tumor identification in lung CT images. These metrics are crucial for understanding the model's strengths and limitations and for comparing its performance to other state-of-the-art approaches. Additionally, we might have also visualized the model's predictions to gain insights into its behavior and potential areas for improvement.

3.4 Experimental Results

CT images play a crucial role in identifying tumors on lung. However, acquiring a large and diverse dataset of real LUNA16 images can be challenging due to various factors such as patient privacy, cost, and limited availability. To address this issue and enhance the training data for identify tumors on lung CT images, YOLO v5x, a state-of-the-art object detection, is utilized to identify tumors from available lung CT images. The process starts by collecting a set of CT images with corresponding lung tumor annotations. These CT images are then used as the input for our model, a network specifically designed for medical image synthesis. Our model consists of a detection network and a classification network, which are trained to Identify tumors on lung CT images. Once our model is trained, it starts identify tumors from the input CT images. The Identified images aim to be distinguishable from real CT images and should capture the essential position required for accurate lung tumor identification.

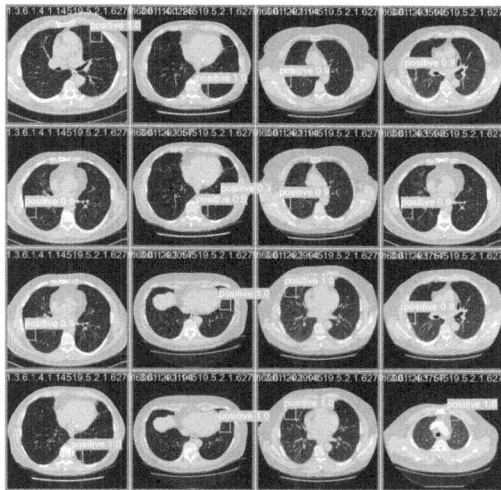

Fig. 4. Experiment to identify tumors on lung CT images

The results shown in Fig. 4 provide valuable insights into the performance of the deep learning model after 100 epochs of training for tumor identification in lung CT images. The identified images exhibit numerous areas with correctly identified tumors. This indicates that the deep learning model is successfully detecting and localizing tumors in the lung CT scans, which is the primary objective of our model. The ability to identify tumors accurately is crucial for early detection and subsequent medical interventions. The explanation for the presence of correctly identified tumors in the source CT images can be attributed to the nature of the training data. During the training process, the model was exposed to a large dataset of CT images that were annotated with lung tumor labels. This annotated data allowed the model to learn the visual patterns and features associated with tumors, enabling it to recognize similar patterns in unseen CT scans. The experimental results indicate that the deep learning model achieves convergence relatively quickly, typically within 100 epochs. Convergence refers to the point in training where the model stabilizes and the learning process reaches a state of equilibrium. In this state, the identified images closely resemble real CT images in terms of quality and appearance, showcasing the model's ability to generalize well to unseen data. As the training progresses, the identified images gradually improve in quality and appearance. This is a positive sign as it indicates that the model is continuously refining its understanding of tumor characteristics and is capable of making more accurate predictions. The ability to generalize to new and unseen CT images is crucial for the model's practical application in real-world medical settings. The results demonstrate that the deep learning model is effective in tumor detection and identification on lung CT images. The model's ability to quickly converge and produce high-quality results speaks to its efficiency and reliability as part of the CAD system. These findings are promising and contribute to the advancements in medical image analysis, particularly in the early detection and treatment of lung tumors, which can lead to improved patient outcomes. However, further evaluation and validation on larger and diverse datasets may be necessary to solidify the model's performance and generalization capabilities.

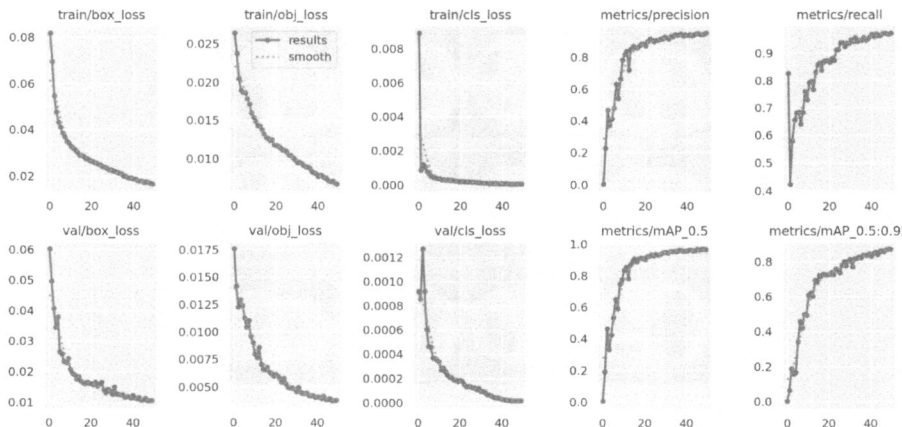

Fig. 5. Progress of train with loss function in identifying tumors on lung CT images with LUNA16 dataset.

The graph in Fig. 5 presents the loss function values of the validation set and the training set during the training process. It shows that the loss function of the validation set converges quite well without significant divergence or overfitting compared to the loss function of the training set. This observation is crucial as it indicates that the model generalizes well to data, and the performance on the validation set is not compromised due to overfitting on the training data. The success of the model in generalizing can be explained by considering several key factors in the data and the training process. The LUNA16 database used for training and validation is characterized by a substantial amount of lung CT images, providing a rich and diverse dataset for the model to learn from. Having a large database helps the model to capture the complexity and variety of lung CT images, improving its ability to generalize to new. The images in the LUNA16 database are in good shape, meaning they are well-annotated, have clear lung boundaries, and exhibit minimal artifacts or distortions. High-quality images contribute to more reliable training and allow the model to learn accurate representations of the lung structures. The synthetic identified images generated by our model are of pretty good quality, as mentioned earlier. These identified images serve as additional data for training and enrich the dataset, enhancing the diversity and increasing the effective size of the dataset. The good quality of these identified images ensures that they contribute valuable information during training, allowing the model to learn relevant features for identifying tumors on lung CT images. By using the identified images generated by our model to enrich the dataset, the overall quantity of available training data is increased. More data means the model has access to a broader set of examples, which aids in building a robust and generalized representation of lung tumor annotation patterns. The combination of a large and high-quality database, along with the introduction of synthetic lung CT images for data enrichment, results in a more effective level of learning. The model can leverage the rich information. It is worth noting that while the current approach has shown promise, additional analysis and evaluation on different datasets or in a cross-validation setting could further validate the effectiveness of the model and its generalization capabilities. Nonetheless, the results presented in Fig. 5 suggest that the proposed method is on the right track in achieving accurate lung tumors identification without compromising generalization performance. The utilization of GIoU (Generalized Intersection over Union) loss function during the training of YOLO v5x plays a crucial role in achieving excellent performance in detecting nodules on lung CT images. GIoU loss is an improved version of the traditional Intersection over Union (IoU) loss, which takes into account the overlap between predicted and ground-truth bounding boxes. By utilizing GIoU loss, the model can better capture the spatial relationships between the predicted bounding boxes and the actual nodules, leading to improved accuracy. To provide a comprehensive assessment of the superiority of the proposed model, we conducted a rigorous comparative analysis with the previously recognized advanced lung nodule detection method, as presented in Table 1. This comparison is intended to introduce the advantages and advancements achieved by our model in terms of efficiency, accuracy, and computational complexity.

Table 1. Compare with various methods

Method	Precision	Recall
Swetha Subramanian [8]	0.893	0.712
Ours	0.947	0.975

In addition, the faster inference time in our model allows for rapid and real-time tumor identification, which is important in time-sensitive medical situations where decisions Prompt identification and prompt patient care are of the utmost importance. The enhanced speed of our model makes it more suitable for use in large-scale lung cancer screening programs where a significant number of CT scans are efficiently processed is essential for early tumor detection and improving public health outcomes.

4   Conclusions

In our research paper, the primary objective is to develop an efficient and accurate model for detecting nodules and distinguishing between nodular and non-nodular regions in lung CT images. To achieve this, we leverage the YOLO (You Only Look Once) v5x model, which is a state-of-the-art object detection algorithm known for its speed and accuracy. Our model is trained using the LUNA16 dataset, which is a well-established and widely used dataset for lung nodule detection and evaluation. The model's performance is assessed using standard evaluation metrics, including precision, recall, and F1-score. The experimental results demonstrate the effectiveness of our proposed model. The evaluation metrics on the LUNA16 dataset show a precision of 94,7%, and recall of 97,5%. These high values indicate that our model is capable of accurately identifying the locations of nodules while maintaining a good balance between precision (low false positives) and recall (low false negatives). The improved efficiency of our model is attributed to the streamlined architecture of our model, which allows for real-time inference without compromising accuracy. The model's lightweight design and efficient use of resources make it a practical and viable solution for lung nodule detection tasks, especially in scenarios where real-time or near-real-time processing is required. Our research demonstrates the effectiveness and efficiency of our proposed model with GIoU loss for lung nodule detection and classification. The high precision, recall, and F1-score obtained on the LUNA16 dataset showcase the model's capability to accurately locate and differentiate between nodular and non-nodular regions in lung CT images. Additionally, the comparison with the previous state-of-the-art model confirms the superiority of our approach in terms of efficiency, making it a promising contribution to the field of lung nodule detection in medical imaging.

References

1. YOLO 5 Homepage. https://github.com/ultralytics/yolov5. Accessed 19 Apr 2023
2. LUNA16 Homepage. https://luna16.grand-challenge.org/ Accessed 09 Apr 2023
3. Rezatofighi, H., Tsoi, N., Gwak, J., Sadeghian, A., Reid, I., Savarese, S.: Generalized intersection over union: a metric and a loss for bounding box regression. In: Proceedings of the IEEE/CVF Conference on Computer Vision and Pattern Recognition, Long Beach, CA, USA, 15–20 June, pp. 658–666 (2019).
4. Lung Cancer Statistic Homepage. https://www.wcrf.org/cancer-trends/lung-cancer-statistics. Accessed 09 Apr 2023
5. Bhandary, A., Ananth Prabhu, G., Basthikodi, M., Chaitra, K M.: Early diagnosis of lung cancer using computer aided detection via lung segmentation approach. Int. J. Eng. Trends and Technol. (2021)
6. Bhattacharjee, A., Majumder, S.: Automated computer-aided lung cancer detection system. In: Advances in Communication, Devices and Networking. Lecture Notes in Electrical Engineering. pp. 425–433 (2019)
7. Zhang, D., Hu, J., Li, F., Ding, X., Sangaiah, A.K., Sheng, V.S.: Small Object Detection via Precise Region-Based Fully Convolutional Networks. Comput. Mater. Continua. (2021)
8. Swetha Subramanian Homepage. https://git.io/Jf9Og. Accessed 19 Apr 2023
9. Xie, H., Yang, D., Sun, N., Chen, Z., Zhang, Y.: Automated pulmonary nodule detection in CT images using deep convolutional neural networks. Comput. Sci. Pattern Recognit. (2019)
10. Ren, S., et al.: Faster R-CNN: Towards real-time object detection with region proposal networks. IEEE Trans. Pattern Anal. Mach. Intell. (2017)

A Context-Aware Application to Monitor the Air Quality

Giacomo Cabri$^{(\boxtimes)}$ and Gabriele Nocetti

University of Modena and Reggio Emilia, Modena 41125, Italy
giacomo.cabri@unimore.it, 270435@studenti.unimore.it

Abstract. The widespread use of mobile devices allows developers to reach and inform more users than ever before through apps. This work presents the design and development of an application that allows anyone to monitor air quality levels, including pollen and pollution. The developed application retrieves official data and uses elements of context awareness to customize the user experience based on the *user's sensitivity* to specific particles. During the development process, special attention will be paid to interfaces and data presentation based on user characteristics and preferences. The project will focus on the Italian territory. Due to the richness in terms of biodiversity of the country, similar solutions exist only at the regional level without taking into account the user's sensibility. Our contribution is to show an application that join general purpose with context-awareness aspects.

Keywords: Pollen · Context-awareness

1 Introduction

Airborne particles, specifically *pollen-like* particles, can cause *allergic symptoms* in sensitive subjects. There appears to be a relationship, currently under study, between the level of concentration of these particles in the air and the symptoms experienced by individuals [2, 8, 13, 15]. This paper is not going to approach the stage of study or analysis to obtain future predictions of these phenomena. Instead, the intent is to develop a *tool* that, having reliable data and predictions available, provides an opportunity for anyone to learn about this topic.

Taking advantage of the diffusion of smartphones, in this paper we present the design and development of a context-aware *mobile application*. Choosing this development target opens up many possibilities for gathering information to integrate a *context-aware* system. The goal is to combine the world of air quality, in order to inform users interested in this topic, with the opportunities offered by the development of mobile platforms to customize the experience.

An application covering this topic (air quality) can certainly lead to the implementation of features to provide benefits for the user, such as allowing him or her to monitor levels of particles in the air and to receive alerts if changes are expected in the coming days. Applying *context awareness* to the alerting

© ICST Institute for Computer Sciences, Social Informatics and Telecommunications Engineering 2024
Published by Springer Nature Switzerland AG 2024. All Rights Reserved
P. Cong Vinh and N. Thanh Tung (Eds.): ICCASA 2023, LNICST 579, pp. 175–185, 2024.
https://doi.org/10.1007/978-3-031-58878-5_13

functionality can *change* its behavior and operation, to the advantage of the user with *notifications* of information that may actually be useful.

The contribution of this paper is to show the design and the implementation of an application that from the one hand is general and can be applied in different areas, but at the other hand it is based on context-awareness, which makes it adaptable to specific situations.

The rest of the paper is organized as follows. After discussing some related work (Sect. 2), we will propose the requirements of our application (Sect. 3). Then, we will present the implementation (Sect. 4), detailing the user interface (Sect. 5). Finally, we conclude the paper with some remarks and future work (Sect. 6).

2   Related Work

In the context of air quality monitoring applications, we consider smartphone apps that can be installed by anyone and that provides data and/or predictions of the particles present in the air of a specific area.

The first example [9] is only available in the Austrian territory and, thanks to the improved *data prediction* model, has also allowed the creation of a personalized real-time experience based on the user's symptom registration.

Another system, called RealForAll [16], uses an *AI-based* subsystem to achieve independence of *aerobiology* experts and allows the use of a symptom calendar for easy comparison of measurements.

Other examples are available for everyone in the apps markets (*App Store & Play Store*), and according to [17] they may have disadvantages. Many of these apps, covering vast territories or even the whole world, have the risk of presenting, in a modern and attractive interface, approximate data that are not specific to a particular area. The problem of the correctness of the collected data must be addressed and one should try not to simplify them during the presentation phase.

In these applications, common *context-awareness* elements can be found:

- presenting the user with data retrieved as close as possible to their *location*;
- providing the ability to *indicate*, on a day-to-day basis, what symptoms the user is experiencing or what pollen particles is *sensitive* to.

Solutions related to the chosen territory, Italy, are often limited to local or, at most, regional areas due to the biodiversity of the area in question. Moreover, mobile applications dealing with these territories currently lack context-awareness elements to help the user visualize the data, something that are present in more modern applications that cover larger territories. In conclusion, solutions that are limited to work in one or more countries are the most effective and accurate because of the data provided by national services. In contrast, those that cover the whole world offer data covering a wider area but with lower precision.

Our proposal will focus on the *Italian territory*, using the national data offered. Therefore, our goal is to provide a *ready-to-use* application with context-awareness elements, working with the data obtained. Nevertheless, the proposed model of application aims to be general and can be applied to other countries.

3 Application Requirements

This section presents the functionalities and requirements necessary to ensure the proper operation of the application. After a general explanation of the existing functionalities, a focus will be placed on the elements of context awareness present.

3.1 Features of the App

The application consists of several *screens*, the most important of which is the home screen. This main page is designed to present the user with only the information they need at that moment, allowing them to explore additional fields by navigating through additional screens. Secondary screens include a page that groups particles of a specific type, followed by a study page that presents data and information related only to the particle being examined. The user can select a particle monitoring station of his or her choice through a search screen to view the data obtained from it. The screen that allows the user to report any symptoms caused by particles in the air and the settings screen complete the list.

Regarding the types of particles presented, in addition to those belonging to the common group called "pollen", it was decided to include *pollutant particles*. With the aim of providing a comprehensive picture of air quality and preventing possible allergic effects caused by it [1], pollution levels will be included as types of particles.

Many of the screens described above are necessary to present the user with the key functionalities of the project. They are listed below in order of *dependency*:

1. visualization of the latest available data;
2. calculation of the not available particle value forecasts for the next few days;
3. geolocalization of the user to provide information based on their location;
4. navigation between available monitoring stations;
5. diary, providing the user with the ability to report particles to which they are most sensitive;
6. sending notifications about changes in particle values.

3.2 Context Awareness Structure

Data that define the context of the application can be divided according to how it is collected. The first type covers data derived from device information, such as time, day, and location. The second includes the user's personal data, which can only be obtained through the user's interaction with a screen.

In Fig. 1, it is possible to see how these data is structured to enable the implementation of the context aware service. One can see the requirements for gaining the data and their strict dependence on it to achieve the end result (providing a personalized notification to the user). It is therefore clear that the user's personal status data *depend* on particle level information, which in turn is retrieved through the location of the device.

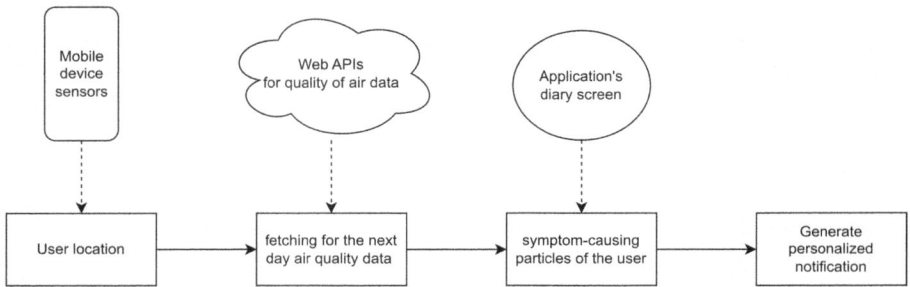

Fig. 1. Context-Aware components

3.3   Notification System

This context-based system will allow the user to receive a maximum of one personalized notification per day. When the application detects changes (increases or decreases) in the levels of the most prevalent particles in today's and tomorrow's data at the last recorded location, the notification is sent. Related to this are two concepts that need to be explained and defined to understand how the system works.

It was decided to *group* particles belonging to a common family into types: spores, pollutants, tree pollen and grass pollen. This choice is not intended to replace the level representation of each particle present, listed by the user, but to facilitate the visualization of information. The subdivision, also applied at the structural level, will condition all the functionality presented, including the context-based notification system.

Each particle is associated with a daily value and a set of *threshold values* that will determine the level reached by that particle on that day. Chosen as absent, low, medium and high, once all levels of a type are collected, the average level of each type will be obtained. In connection with the notification system, the way to determine whether to send a notification is based on the presence of changes in the values belonging to the mean or high levels of the most prevalent type on the current or next day.

At this point, the integration of the user's personal data into the notification system occurs in two stages. The *first* stage involves data collection, which is done through the *symptom report page* of the application. Once a day, the user can fill in the screen by entering the number of hours spent outdoors and the intensity of symptoms felt. This data will be processed along with particles belonging to the most prevalent type detected that day. At the end of this stage, a *weight* is assigned to the detected particles (which may have caused those symptoms). During the *second* stage, when the notifications are generated, these weights will be applied when comparing particle levels. The ways in which they may affect will be explained in the following sections.

The notification system includes an additional notification, concerning only changes in *pollutant levels*, independent of personal data. In conclusion, the system is working from the first app-launch, but will be customized once the user interacts with the diary screen.

4 Application Implementation

After an introduction regarding the framework used for development, the implementation of the features in the application will be broken down and described below.

4.1 The Chosen Framework

The *Flutter framework* was selected for the implementation of all application features. Focusing on the development of mobile platforms for *Android*, the "material" library [5] included in the framework made it possible to create interfaces that were *OS-consistent* and easy to use.

The framework is based on the *Dart* programming language, which is oriented toward web and mobile development and places great emphasis on handling asynchronous operations and *Future* data types, suitable for the structure of the chosen application.

4.2 Data Retrieval

The sources selected, based on the requirement of data reliability, are two. *POLL-net* [14], the Italian aerobiological monitoring network of the National System for Environmental Protection (SNPA), for pollen and spores. The open source *Open-Meteo* [11] project concerning air pollutants. The last one allows interfacing with data from the Copernicus Atmosphere Monitoring Service (CAMS) [3]. Both of these sources will be queried through their *APIs*, returning *JSON data* processed with *Dart*. The threshold values for particle levels were retrieved from the POLLnet service and the guidelines [12] of the World Health Organization (for air pollutants).

As noted in Sect. 2, the basic requirement for the data used in this context is *reliability*. This requirement is luckily met, in the borders of *Italy*, by the POLLnet service. When delineating the quality of the data, which includes both the diversity of pollen types identified and the additional information attributes, it becomes evident that by restricting the scope to the Italian territory, a degree of *accuracy* quite different from that of other international services is achieved.

Among these considerations, there is also to be added the *amount* of monitoring stations located across the various regions of Italy. Taken together, these make up a total of 64 stations. This collective infrastructure, combined with the rich quality of the data, guarantees end users of the application (which is based on this network) a *remarkable* standard of accuracy and data quality.

It is important to emphasize, as mentioned in the abstract, the fact that we work only with data obtained from official sources. Among these, the most important ones are those related to the forecast of particle levels in the coming days. For air pollutants, CAMS provides readily available and already calculated data. However, getting data predictions is different for those concerning pollen and spores, which are obtained from POLLnet. Since POLLnet provides data only up to the previous week, we needed to find a way to calculate data for the following days. This was done following guidelines [7], but it is only meant to be a temporary solution until more reliable data regarding pollen and spores forecasts are obtained.

After obtaining the device location, data are *retrieved* from Open-Meteo using latitude and longitude data. Retrieving data from POLLnet is more challenging, as it requires determining the closest monitoring station to the device location using the *Haversine formula*, which calculates the distance between two geographic coordinates.

To ensure a satisfying *user experience* in different situations, especially considering that the application is intended for mobile devices, two specific use cases in particular were covered. First, in case there is no GPS signal, the system ensures that users have access to data from all monitoring stations, in *browse-mode*. Second, even when GPS functionality is active, if the nearest monitoring station has no data available, the app will show information from the nearest *working* monitoring station to users.

4.3   Presentation Layer

In terms of data visualization, the user has access not only to the "Home" summary screen, but also to the tab for each detected particle. This tab features a *graph*, drawn using external library[1], that shows the user the latest official data retrieved based on the chosen monitoring location.

Each screen of the application presents particles categorized by type and sorted by intensity level. In order to achieve a *common view* of these screens, it was decided not to apply differences in the sorting depending on the sensitivity of the user regarding certain particles. Screenshots of the application pages are available in the following sections.

---

[1] https://pub.dev/packages/fl_chart.

4.4   Storing the Data

External plugins can be used in Flutter to provide data storage services. The first plugin[2] uses Android's "Shared Preferences" to store *non-sensitive* information, such as the flag that enables or disables notifications. To speed up the application and reduce the number of calls to external sources, a *caching service* was implemented using another plugin[3]. Finally, a plugin that prioritizes *security* was chosen to store the most sensitive data, such as the user's last detected location and the particles to which the user is most sensitive. This plugin[4] works based on the *KeyStore system* when in the *Android* environment.

5   An Overview of the User Interface

5.1   Homepage and Details of Particles

Every time the application starts, the home screen is presented. Depending on the location of the device, it shows the highest-level particle type in the foreground. The screen is divided into *tabs* representing today and the following days, and is enhanced by a small box showing the *weather values* for the selected area.

Each type of particle can be viewed in detail on its specific page, which also shows the *distance* and accuracy of the data by location. In addition, a *detailed tab* for each individual particle can be viewed, with a corresponding graph. This tab shows the data source, threshold values for the levels, and a brief description. In the upper right corner, the search icon allows users to search for monitoring stations and view data for that location. The button in the lower right corner provides access to the screen for *reporting* any symptoms caused by particles in the air. It is only visible on the home screen and becomes available again the next day after being used.

All the screens just described can be viewed in Fig. 2, and all the *logos* representing the types were generated by Midjourney [10].

5.2   Symptom Report Page and Diary

Thanks to the screen showed in Fig. 3 left, the user can choose the level of *reported symptoms* and the number of *hours spent outdoors*. When finished, the process explained in Sect. 3.3 is performed and the data are saved. Only data on particles exceeding a minimum level of presence in the air are saved. For demonstration purposes, the medium level was chosen.

These data, sorted from the most recent, can be viewed in the diary screen from the settings menu. Protected by *biometric* authentication (fingerprint), the data can be viewed and, if necessary, deleted.

---

[2] https://pub.dev/packages/shared_preferences.
[3] https://pub.dev/packages/flutter_cache_manager.
[4] https://pub.dev/packages/flutter_secure_storage.

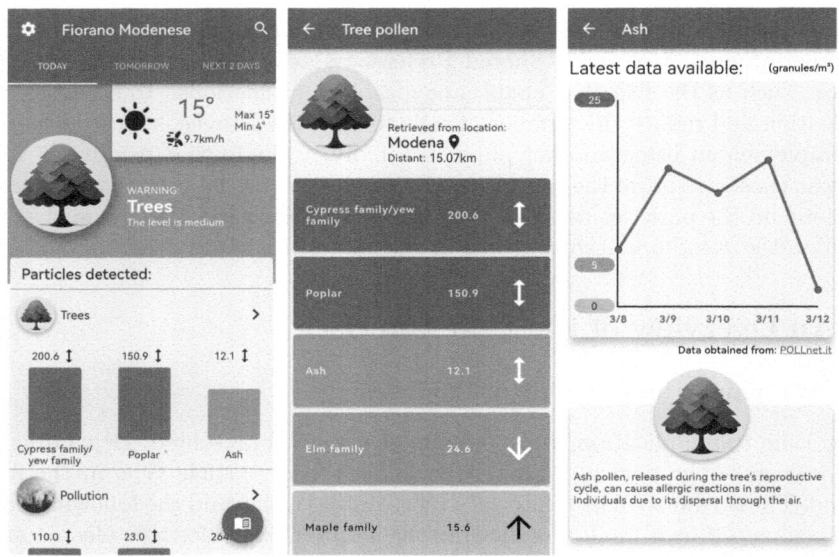

Fig. 2. Homepage (left), view of the type (center) and particle details (right)

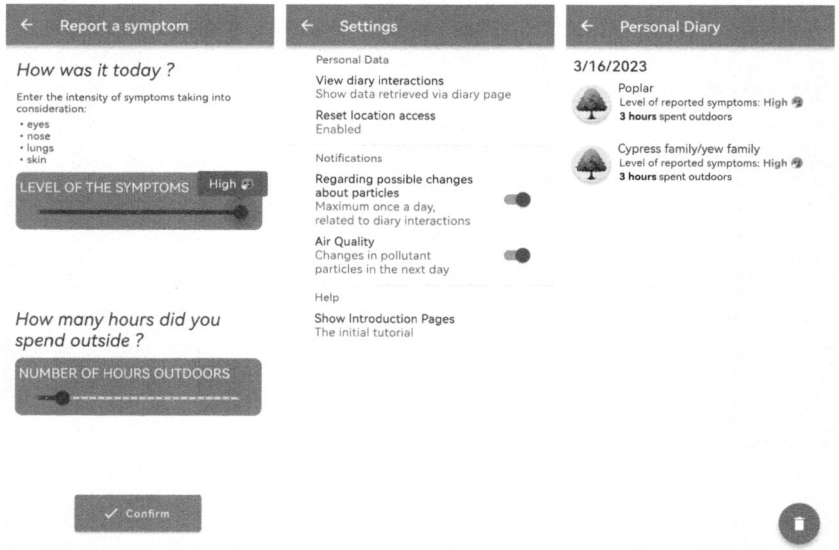

Fig. 3. Symptom reporting page (left), settings (center) and diary screen (right)

5.3   Notifications

Notifications are customized according to the last recorded location and particles saved in the *diary*. This may be clearer after an example.

The device is located in Treviso (Fig. 4), where the level of polluting particles is higher than tree pollen. A notification is displayed, warning of increased pollution, because the user has enabled air quality notifications. In addition to this, it is noted that the level related to the *cypress family* is also increasing, and in this case there are 2 possible scenarios:

1. if the diary is *empty*, priority will be given to the pollution type;
2. if the diary has *marked* the cypress family enough times, the notification will alert the sensitive user about the increase of that particle.

In the second scenario, the diary is filled with the values in Fig. 3. The high level of symptoms presented, combined with the number of hours spent outdoors, gave the particles of the cypress family *more weight* than the increase in pollution. In addition to these exemplary cases, it was essential to ensure the perfect functionality of the custom notification distribution logic in a wide range of scenarios.

For this purpose, along with the development of the necessary operating code, an exhaustive set of *unit tests* [6] was made, examining the behavior of the notification system in every possible permutation. This testing includes a total of 256 distinct *configurations* based on the different *intensity ranges* that 8 particles of 2 different pollen types can manifest on a given day (i.e., absent, low, medium, high). The tests also include the interaction of 8 potential "weights" values of the particles, generated by interactions with the *journal* screen, all of which are accurately stored on the device.

The weight combinations subjected to testing includes values associated to two distinct types of pollen particles, in order to introduce a *dynamic* element that influences the behavior of the notification system. These weights are systematically grouped across the *full spectrum* of possible pollen intensity combinations, producing a complete examination of system *responsiveness*.

Together with other *practical* tests conducted on devices of different brands, all of these tests played a key role in the development, significantly contributing to the refinement and optimization of the notification system.

5.4   Testing on Different Devices

The application was developed to work on screens with resolution higher than 320×480 px. The interface was tested on devices belonging to all major Android smartphone manufacturers and did not present any particular problems.

The notification system, specifically when *showing* notifications, presented different behaviors depending on the device. From these tests it appears that on stock Android systems, without any special modifications by the producer, notifications are displayed almost all the time. The more the producer customizes

Fig. 4. Particle levels today (left) and tomorrow (center) in Treviso, notification when diary is empty (upper right) and when cypress family is in the diary (lower right)

her Android system in the area of background app management, the less notifications are likely to be displayed often. To solve this problem, it is often necessary for the user to change the system settings that regulate the background app's behavior. The results collected from the tests turn out to be very similar to those of the "Don't Kill My App" project [4].

6   Conclusions

In conclusion, this paper describes an *Android-based* application that allows a large number of users to *monitor* particle levels in the air over the Italian territory. A *context-aware notification system* has also been implemented, which customizes the user experience according to the user's sensitivity to certain particles. The *innovative* approach of this system is highly appreciated as it is applied in a field of great interest, such as the health of ordinary citizens. Even more so in the area of Italy, where solutions of this type are only present at a regional level and lack context-awareness elements.

Thanks to the *cross-platform* framework on which the application is based, additional work in the near future includes support for Web, Apple, and Windows devices. Currently, the project can be compiled in JavaScript and run on any browser without modification. As a result, *platform-specific* functions, such as notifications and location permissions, will need to be converted to the new ones.

A further step would be to expand the service to operate *globally*, leveraging the CAMS pollution level retrieval service, which already offers data covering the whole world. This more complex future development will have to follow the same logic chosen for the Italian territory, relying on official data from the relevant county where it is possible.

However, the most important upcoming work will be to analyze the functionality of the application with the support of *aerobiology experts* and ensure that it is not only technologically reliable, as demonstrated, but also scientifically valid.

References

1. Bartra Tomàs, J., et al: Air pollution and allergens. J. Investig. Allergol. Clin. Immunol. **17**, supl. 2, 3–8 (2007)
2. Brito, F.F., et al.: Olea Europaea pollen counts and aeroallergen levels predict clinical symptoms in patients allergic to olive pollen. Ann. Allergy Asthma Immunol. **106**(2), 146–152 (2011)
3. Copernicus - C.A.M.S. https://atmosphere.copernicus.eu/. Accessed 14 Mar 2023
4. Don't kill my app project. https://dontkillmyapp.com/. Accessed 16 Mar 2023
5. material library - Dart API. https://api.flutter.dev/flutter/material/material-library.html. Accessed 14 Mar 2023
6. Online repository of the project. https://github.com/noceg43/pollen_app. Accessed 21 Sept 2023
7. Guidelines from POLLnet (in Italian). https://www.isprambiente.gov.it/files2017/pubblicazioni/manuali-linee-guida/445364_Manuale_linee_guida_151_17.pdf. Accessed 14 Mar 2023
8. Jantunen, J., Saarinen, K., Rantio-Lehtimäki, A.: Allergy symptoms in relation to alder and birch pollen concentrations in Finland. Aerobiologia **28**, 169–176 (2012)
9. Kmenta, M., Bastl, K., Jäger, S., Berger, U.: Development of personal pollen information-the next generation of pollen information and a step forward for hay fever sufferers. Int. J. Biometeorol. **58**, 1721–1726 (2014)
10. Midjourney website. https://www.midjourney.com/. Accessed 17 Mar 2023
11. Free Open-Source Weather API — OpenMeteo. https://open-meteo.com/. Accessed 14 Mar 2023
12. Organization, W.H., et al.: Who global air quality guidelines: particulate matter (pm2. 5 and pm10), ozone, nitrogen dioxide, sulfur dioxide and carbon monoxide: executive summary. WHO global air quality guidelines: particulate matter (PM2.5 and PM10), ozone, nitrogen dioxide, sulfur dioxide and carbon monoxide (2021)
13. Rapiejko, P., Stanlaewicz, W., Szczygielski, K., Jurkiewicz, D.: Threshold pollen count necessary to evoke allergic symptoms. Otolaryngologia Polska= The Polish Otolaryngol. **61**(4), 591–594 (2007)
14. SNPA - Sistema nazionale protezione ambiente (in italian). https://www.snpambiente.it/dati/pollini-e-spore-fungine/. Accessed 14 Mar 2023
15. Tegart, L.J., et al.: 'pollen potency': the relationship between atmospheric pollen counts and allergen exposure. Aerobiologia **37**(4), 825–841 (2021)
16. Tešendić, D., et al.: Realforall: real-time system for automatic detection of airborne pollen. Enterprise Inform. Syst. **16**(5), 1793391 (2022)
17. Travaglini, A., et al.: Approximate or accurate? efficacy of daily use of weather and air quality mobile applications for pollen allergy sufferers? Pediatric Allergy Immunol. **33**, 41–43 (2022)

Applying Guided Discovery Learning to Enhance the Achievement of Information Technology Team

Cam Ngoc Thi Huynh[1,2], Anh Van Thi Tran[3], Tha Thi Bui[4], Hong Thi Nguyen[5], and Phuoc Vinh Tran[1,5(✉)]

[1] Institute of Applied Mechanics and Informatics, Ho Chi Minh City, Vietnam
camhuynhit@gmail.com, phuoc.gis@gmail.com
[2] Graduate University of Science and Technology, Ha Noi, Vietnam
[3] Hochiminh College of Economics, Ho Chi Minh City, Vietnam
anhttv@kthcm.edu.vn
[4] University of Labour and Social Affairs (Campus II), Ho Chi Minh City, Vietnam
thabt@ldxh.edu.vn
[5] Thudaumot University (TDMU), Binhduong, Vietnam
{hongnt.ktcn,phuoc.tv}@tdmu.edu.vn

Abstract. The fast development of IoT and ChatGPT is urging the change of educational method. The traditional methods are being gradually replaced with discovery learning, the method of teacher-centered learning is transferred to student-centered learning. The discovery learning method is developed to guided discovery learning model to apply in educational institutions. This experimental research applied guided discovery learning method for training the information technology team of a gifted high school to take national excellent student prize. A clustering algorithm is applied to set up the team. The research also constitutes an algorithm to evaluate the efficiency of guided discovery learning method for each lecture, where teacher is involved in the evaluation. The result of guided discovery learning method is proved with 3 prizes got by the 2022 team comparing with 1 prize of the 2021 team applying traditional learning method, which is considered as control team of this experiment. An issue which need to be discussed is the increase of teacher's time and energy for the preparation before class and teacher's self-motivation during class to apply guided discovery learning method.

Keywords: Discovery Learning · Guided Discovery Learning · National Excellent Student

1 Introduction

The rapid development of data sources on internet and ChatGPT has been urging educational institutions to change educational method, from centered-teacher learning to centered-student learning. For the methods of centered-student learning [1], several

P. Cong Vinh and N. Thanh Tung (Eds.): ICCASA 2023, LNICST 579, pp. 186–196, 2024.
https://doi.org/10.1007/978-3-031-58878-5_14

institutions applies discovery learning, where student is the center of all learning and teaching activities. So far, the model of discovery learning for educational institutions is still being discussed, some authors are arguing about the role of teacher in discovery learning model, discovery learning with teacher or without teacher.

In a traditional class, the teacher is not necessary to completely cognize in order to stimulate students' skills of creativity, criticalness as well as their self-motivation in learning. The lecture is still designed according to the purpose of curriculum and the expectation of teacher. It is really difficult for teacher to cognize students' skills and attitude because of their variety in thinking and self-motivation. The problem to be solved is how to apply the discovery learning model with centered-student and guiding-teacher for an educational institution.

This research is an experiment applying guided discovery learning method in a gifted high school for the team in information technology. The research determines the teacher's functions in applying discovery learning process. The research applies a clustering algorithm to choose some students of the similarly good competence into the team. The research constitutes the algorithm to evaluate the efficiency of leaning and teaching activities after each lesson, where teacher's competence is involved as a variable of the algorithm.

The article is structured as follows. The Sect. 1 introduces the trend the development of the discovery learning in the era of internet, IoT, and ChatGPT. The Sect. 2 presents guided discovery learning method with teacher's functions in discovery learning process. The Sect. 3 experiments guided discovery learning with the information technology team of a gifted high school to take national excellent student prize; this section applies a clustering algorithm to set up the team and constitutes the algorithm evaluating the efficiency of guided discovery learning process. The Sect. 4 presents the results of the experiment by comparing with the team of the last year, and some issues to discuss. Finally, the conclusion summarizes the results of the research.

2 Guided Discovery Learning Model

2.1 Discovery Learning

Since early 21st century, active learning has emerged as new trend of educational systems. Indeed, several authors are arguing that the learning method of students at educational institutions is nowadays developing from expository to discovery learning [2, 3], in which the central position of educational activities is transferred from teacher to student [4] to promote students' role in learning activities [5]. As an active learning approach, discovery learning with centered-student contrasts with the traditional learning and is superior to the traditional learning with centered-teacher and passive students [1, 2].

Discovery learning is a new educational model, where students' ability and self-reliance are promoted to self-find and construct new knowledge easy to long store in memory [6, 7]. Some authors consider that the basic characteristics of discovery learning model are active learning, meaningful learning, self-efficacy [3, 8–10], meanwhile, Piaget's theory of cognitive development considers that students cannot self-process and self-understand information which they receive [11], hence some educational institutions are developing the guided discovery learning approach [12–14].

2.2 Guided Discovery Learning

For discovery learning or guided discovery learning, each student has to self-discover the concepts or rules from collected data to acquire into long-term memory as new private knowledge [15]. For guided discovery learning in educational institutions, teacher guides the students of a class to identify learning topic and problems, discover concepts and/or rules to acquire into long-term memory as new knowledge. Teacher applies the graph theory [16–18] and visualization techniques [19, 20] to design the lectures for guiding students to carry out process of discovery learning [4, 21–23].

• *Step 1: Stimulating activity and thinking*

The visual lecture designed by teacher applies visual graphs to attract students' interest and attention in learning activities and motivate their self-reliance as well as thinking to focus on the academic field of topic.

• *Step 2: Identifying topic and expressing problems*

Identifying topic and expressing problems is an important step, the visual lecture shows students some literatures to evoke some similar topics, e.g. the topic of hand, foot, mouth disease [9, 10], the topic of dengue fever [24], etc. Each student can state his thinking to be feedbacked by the teacher about the topic which he expresses. After accepting a topic for the whole class, the teacher utilizes visual graphs to continuously evokes students many problems, e.g. the development of epidemic in the year of 2021? [24], the dangers happening disease? [9, 10], etc. After that, the teacher guides students to express problems along with the purpose accordant with the topic.

• *Step 3: Collecting data*

The visual lecture evokes students the models of data tables and shows them data sources related to each variable. The teacher guides students how to collect data from the various sources and fill appropriate data in the chosen tables. Data can be collected by accessing internet, by reading literature, or by interviewing related persons, etc… The visual lecture can evoke students the way to cluster and arrange data according to variables and their relations.

• *Step 4: Analyzing data*

Analyzing data refers to answer questions based on data [10]. The teacher guides students to represent the data tables as visual graphs and creatively make analytical questions composed of local questions, global questions, relative questions [10]. Analytical questions can be answered by visualization or algorithm approach [25]. The answers of analytical questions can result in the rules which students have not known, e.g. the high correlation among rainfall, humidity, and time can happen the danger of hand, foot, mouth disease, meanwhile average temperature does not relate to [10].

• *Step 5: Verifying*

The teacher guides students how to compare the step-4 answers with the purposes of the problems asked at the step 2 and utilize several various data sources to examine the results of data analysis. The teacher guides students to examine the conclusions with several data sources, e.g. it is necessary to have to collect more data in several years to verify the discovery in the example (e.g.) at the step 4 [10]. With the confident data sources, if only an analytical result is not verified, all results are considered as not be verified.

• *Step 6: Extracting conclusions*

If all results of the examination are verified and can be understood as knowledge, they may be generalized as rules to apply for similar problems [25]. The teacher guides students to express the results as conclusions of the topic or the rules for other similar problems. Consequently, the teacher guided students to discover rules as their new knowledge.

3   Research Method

This research is experimented at a gifted high school for training the information technology team taking national excellent student prize. The research applies a clustering approach to group the information technology students of the similarly good competence into the team. The coach is responsible for coordinating training activities which is carried out by several teachers, each guides a topic. After each topic, the leaning and training activity is evaluated by the algorithm assessing efficiency, constituted by the research.

3.1   Grouping Students into Team by Competence

Student's competence is represented as a vector of features which are composed of learning features and non-learning features [26]. Learning features are prior learning outcomes. Non-learning features are composed of skills and attitude as self-motivation, self-confidence, self-reliance, self-finding, self-investigation, self-analysis, critical thinking, creative thinking; and socio-economic factors as demographic, family, behavior backgrounds, interaction, where family background including finance factor and interest is important. The non-learning features are evaluated by pre-tests and interviews of candidates together with their parents.

Input

$X = \{x_n | n = 1, 2, .., N\}$ is the vector set representing the students who are candidates for the team, where $x_n = [s_{1.n}, .., s_{i.n}, .., s_{I.n}]^T = [s_{1.n}; ..; s_{i.n}; ..; s_{I.n}]$ is the feature vector of the student x_n, and $s_{i.n}$ is a feature of learning or non-learning of the student x_n.
$W^{dis} = [w_1^{dis}, .., w_i^{dis}, .., w_I^{dis}]$ is the weight-tuple of competence features defined by each discipline, e.g. information technology.

Output

$Y = \{y_1, .., y_m, .., y_M\}$ is the team composed of $M | M < N$ teamers (team members).

Algorithm

Step 1: Defining the competence features of each candidate $\mathbf{x}_n | n = 1, 2, .., N$ according to the discipline:

$$\mathbf{x}_n^{dis} = W^{dis}\mathbf{x}_n = [w_1^{dis}, .., w_i^{dis}, .., w_I^{dis}][s_{1.n}; ..; s_{i.n}; ..; s_{I.n}]$$
$$\mathbf{x}_n^{dis} = [w_1^{dis}s_{1.n}; ..; w_i^{dis}s_{i.n}; ..; w_I^{dis}s_{I.n}]$$

Step 2: Calculating candidate competence for the discipline

$$\left\| \mathbf{x}_n^{dis} \right\| = \sqrt{\left(w_1^{dis}s_{1.n}\right)^2 + .. + \left(w_i^{dis}s_{i.n}\right)^2 + .. + \left(w_I^{dis}s_{I.n}\right)^2}$$

Step 3: Arranging $\left\| \mathbf{x}_n^{dis} \right\|$ as a descending sequence of N elements, then map them onto $\{\mathbf{y}_1, .., \mathbf{y}_n, .., \mathbf{y}_N\}$, where $\|\mathbf{y}_{n-1}\| \geq \|\mathbf{y}_n\| | n = 2, 3, .., N$, with $\mathbf{y}_n = [s_{1.n}^{dis}; ..; s_{i.n}^{dis}; ..; s_{I.n}^{dis}]$
Step 4: Choosing the segment $\mathbf{Y} = \{\mathbf{y}_1, .., \mathbf{y}_M\}$ as the team.

3.2 Evaluating the Efficiency

The output O refers to the new information and concepts discovered by teamers and the new knowledge learned by teamers. The inputs refer to the brain power which teamers must load and the time which teamers have to take to generate outcomes (Fig. 1). The relation between output and input is mathematically represented by the expression 1.

The input factors which affect teamers' brain load to generate outcomes are composed of the characteristics of lecture as the match H, visualization V, and complexity C; the characteristics of teamers as prior-knowledge K, skill S, and attitude A; and the pedagogic competence of teacher P (Fig. 1), mathematically represented by the expressions 2 and 3.

$$B \times T \rightarrow O \tag{1}$$

$$K \times S \times A \times H \times V \times C \times P \rightarrow B \tag{2}$$

O: Outcome. The outcome refers to the things that the teamers discover. The outcome of each teamer evaluated by the post-test is $o_m | o_m \in \{1, 2, .., 10\}$ for $m = 1,.., M$. The outcome of team is defined as the median or the arithmetic mean of $\{o_m | m = 1, 2, .., M\}$, mathematically $O = median\{o_m | m = 1, 2, .., M\}$ or $O = \left(1/M\right)\sum_{m=1}^{M} o_m$.

B: Brain Load. The brain load is the resources of working memory mobilized by teamers to obtain the outcomes. The lighter the brain is loaded, the higher the efficiency of activity is.

T: Time. The time is the time interval which teamers take to obtain the outcomes. The shorter the time is taken; the higher the efficiency of activity is. The time is considered as a cognitive cost of all activities. Hence, the relation between the variable of time and others is not necessary to study. T is the time of teamers taken by the teacher to guide them by visual lecture according to discovery learning process. The real times are normalized by the range [1, 2,.., 10] for all teamers and all topics.

K: Prior-knowledge. The prior-knowledge refers to available knowledge of teamers before approaching the topic. The more compatible with available knowledge of teamers the lecture is, the lighter the brain is loaded. The prior knowledge of each teamer evaluated by the pre-test (the test before approaching the topic) is $k_m | k_m \in \{1, 2, .., 10\}$ for $m = 1,.., M$. The prior knowledge of team is defined as the median or the arithmetic mean of $\{k_m | m = 1, 2, .., M\}$, mathematically $K = median\{k_m | m = 1, 2, .., M\}$ or $K = \left(1/M\right) \sum_{m=1}^{M} k_m$.

S: Skill. The skill refers to teamer' ability of logical, creative and critical thinking utilized during discovery learning. The easier the skills of thinking are mobilized, the lighter the brain is loaded. The skill of each teamer evaluated by the pre-test is $s_m | s_m \in \{1, 2, .., 10\}$ for $m = 1,.., M$. The skill of team is defined as the median or the arithmetic mean of $\{s_m | m = 1, 2, .., M\}$, mathematically $S = median\{s_m | m = 1, 2, .., M\}$ or $S = \left(1/M\right) \sum_{m=1}^{M} s_m$.

A: Attitude. The attitude refers to teamers' behaviors in learning such as self-motivation, self-confidence, self-reliance, self-finding, self-investigation, self-analysis, and interaction. The more active the attitude is, the lighter the brain is loaded. The attitude of each teamer evaluated by the pre-test is $a_m | a_m \in \{1, 2, .., 10\}$ for $m = 1,.., M$. The attitude of team is defined as the median or the arithmetic mean of $\{a_m | m = 1, 2, .., M\}$, mathematically $A = median\{a_m | m = 1, 2, .., M\}$ or $A = \left(1/M\right) \sum_{m=1}^{M} a_m$.

H: Match. The match refers to the compatibility of visual graphs in lecture with teamers' prior-knowledge schemata. The match of visual lecture assists teamers in carrying out discovery learning process. The higher the match of visual graphs with teamers' prior-knowledge schemata is, the lighter the brain is loaded. The match of visual lecture is evaluated by the teamers' answers to questionnaires, where the arithmetic mean of scores assigned to questions by the teamer m is rounded up or down to $h_m | h_m \in \{1, 2, .., 10\}$ for $m = 1,.., M$. The match of visual lecture evaluated by team is defined as the median or the arithmetic mean of $\{h_m | m = 1, 2, .., M\}$, mathematically $H = median\{h_m | m = 1, 2, .., M\}$ or $H = \left(1/M\right) \sum_{m=1}^{M} h_m$.

V: Visualization. The visualization refers to the visual features of lecture such as beauty, orientation, and stimulation. The higher the visual features are, the lighter the brain is loaded. The complexity generated by visualization is considered as part of the complexity of lecture. The visualization of visual lecture is evaluated by the teamers' answers to questionnaires, where the arithmetic mean of scores assigned to questions by the teamer m is rounded up or down to $v_m | v_m \in \{1, 2, .., 10\}$ for $m = 1,.., M$. The visualization of visual lecture evaluated by team is defined as the median or the arithmetic mean of $\{v_m | m = 1, 2, .., M\}$, mathematically $V = median\{v_m | m = 1, 2, .., M\}$ or $V = \left(1/M\right) \sum_{m=1}^{M} v_m$.

C: Complexity. The complexity refers to the structure of visual lecture, domain, issue, data, visual model, and reference time. The more the complexity is, the heavier the brain is loaded. The complexity of visual lecture is evaluated by the teamers' answers to questionnaires, where the arithmetic mean of scores assigned to questions by the teamer

m is rounded up or down to $c_m | c_m \in \{1, 2, .., 10\}$ for $m = 1,..., M$. The complexity of visual lecture evaluated by team is defined as the median or the arithmetic mean of $\{c_m | m = 1, 2, .., M\}$, mathematically $C = median\{c_m | m = 1, 2, .., M\}$ or $C = \left(\frac{1}{M}\right) \sum_{m=1}^{M} c_m$.

P: Pedagogic Competence. The pedagogic competence refers to teacher's skill applying guided discovery learning method. The higher the teacher's pedagogic competence is, the lighter teamers' brain is loaded. The teacher's pedagogic competence is evaluated by the teamers' answers to questionnaires, where the arithmetic mean of scores assigned to questions by the teamer m is rounded up or down to $p_m | p_m \in \{1, 2, .., 10\}$ for $m = 1,..., M$. The teacher's pedagogic competence evaluated by team is defined as the median or the arithmetic mean of $\{p_m | m = 1, 2, .., M\}$, mathematically $P = median\{p_m | m = 1, 2, .., M\}$ or $P = \left(\frac{1}{M}\right) \sum_{m=1}^{M} p_m$.

The Efficiency. The efficiency of a lecture for guided discovery learning method is defined as the correlation between outcomes and the power of brain which teamers must load, and the time they must take (Fig. 1). The efficiency of the lecture is evaluated by the expressions 4 and 5.

$$B \equiv C \times K^{-1} \times S^{-1} \times A^{-1} \times H^{-1} \times V^{-1} \times P^{-1} = C \times (K \times S \times A \times H \times V \times P)^{-1} \quad (3)$$

$$E \equiv \frac{O}{B \times T} \quad (4)$$

$$E \equiv \frac{O \times K \times S \times A \times H \times V \times P}{C \times T} \quad (5)$$

where the symbol ".. $\equiv$ .." is defined as ".. is directly proportional to..."

In reality, the variables are evaluated for a team of M teamers with each lecture. The outcome O is evaluated by the post-test. The variables of K, S, A are evaluated by the pre-test. The variables H, V, C, and P are evaluated by surveying the teamers. The time T is recorded for each lecture.

Mathematically, the efficiency of a lecture for guided discovery learning method may be quantified from the expression Eq. (5) as follows:

$$E = \frac{O + K + S + A + H + V + P}{C + T} \quad (6)$$

The possible values of E in Eq. (6) from the minimum $\frac{7}{20}$ to the maximum $\frac{70}{2}$ may be normalized from 1% to 100% by multiplying $\frac{O+K+S+A+H+V+P}{C+T}$ and $\frac{20}{7}\%$ to obtain the efficiency of a lecture for guided discovery learning method:

$$E = \frac{O + K + S + A + H + V + P}{C + T} \times \frac{20}{7}\% \quad (7)$$

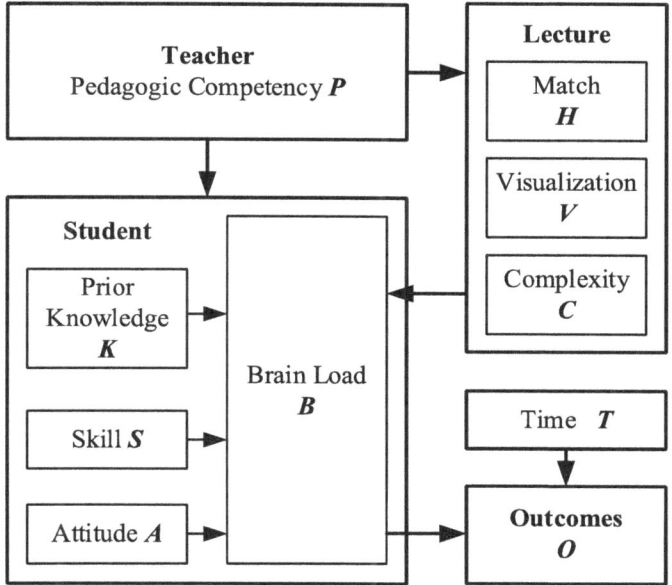

Fig. 1. The model for evaluating the efficiency of a topic in guided discovery learning method.

4   Result and Discussion

4.1   Result

The guided discovery learning method applied for training information technology teamers is estimated by comparing the results of experiment team in 2022 with control team in 2021.

In 2022, the gifted high school welcomes 3 winners from the national excellent student contest in information technology and the remainders got high scores. Meanwhile the 2021 team only got 1 prize in information technology of the contest.

Most teamers of the year 2022 are self-confident and had not lost their self-control while facing exam questions. The losing of self-control of 2021 teamers is one of causes of less result than 2022 team though their knowledge is likely to cover the problems of exam question.

4.2   Discussion

The method of guided discovery learning not only enhances academic knowledge of teamers but also practices their self-reliance and forms their skills to face new problems. This method increases of teamer's self-confidence and avoids to lose one's self-control as facing new topic or new problem.

The method of choosing teamers based on both learning and non-learning competence enhances the effectiveness of cooperative learning in warm discussions to determine topic, expressing problems, collecting and analyzing data, drawing conclusion.

The cooperative learning of teamers of the similar learning and non-learning competence increases teamer's ability facing the stress in contest.

The method assessing the efficiency of each topic encourages teacher in preparing lecture as well as guiding teamers at class because this evaluation involves teacher's working. In addition, this evaluative method assists teamers in awaking to the value of time during taking contest.

The method assessing the efficiency of each topic stimulates teacher's activities in guiding team. Though teacher is not the center in learning and teaching activities of guided discovery learning model, the method of efficiency assessment proves the necessity of teacher's role in the model.

Teachers must invest much time and energy for guided discovery learning method than traditional learning method. Before class time teachers have to set up topics and prepare problems as well as visual lecture to evoke teamers. During class time, teacher must self-motivate to grasp the development of idea of each teamer and guide him.

The experiences from the application of guided discovery learning for this team is being deployed to the information technology classes of the gifted high school. It is necessary to train teachers before extending the guided discovery learning model to the other disciplines of the school.

5   Conclusions

This research aims at experimenting to complete guided discovery leaning model for educational institutions. The experiment was applied to train a team preparing to take national excellent student contest in information technology. Teachers' functions are determined in each step of discovery learning process. The research proposes to apply a clustering approach to select good students for the team and assess guided discovery learning model based on the efficiency of learning and teaching activities of each lecture. Comparing with control team applied traditional learning method of the last year with 1 prize, the experiment team of the year 2022 applied guided discovery learning method wins 3 prizes.

References

1. J. R. Goldberg and M. L. Nagurka, "Enhancing the Engineering Curriculum: Defining Discovery Learning at Marquette University," in The 2012 Frontiers in Education Conference: Soaring to New Heights in Engineering Education, Seattle, Washington, 2012, pp. 405–410: IEEE (2020)
2. Kristiyajati, A., Wijaya, A.: The Effectiveness of Visualization of Proofs in Learning Mathematics by Using Discovery Learning Viewed from Conceptual Understanding. Southeast Asian Mathematics Education Journal 9(1), 37–44 (2019)
3. Svinicki, M.D.: A Theoretical Foundation for Discovery Learning. Adv. Physiol. Educ. 20(1), S4–S7 (1998)
4. T. A. Prasetya, C. T. Harjanto, Improving Learning Activities and Learning Outcomes Using the Discovery Learning Method " Journal of Mechanical Engineering Education, vol. 5, no. 1, pp. 59–66 (2020)

5. Kartika, Y., Hutapea, N.M., Kartini, K.: Mathematical learning development using discovery learning model to improve mathematical understanding skills of students. J. Educ. Sci. **4**(1), 124–132 (2020)

6. Rahman, M.H.: Using discovery learning to encourage creative thinking. Int. J. Soc. Sci. Educ. Stud. **4**(2), 98–103 (2017)

7. M. R. Ramdhani, "Discovery Learning with Scientific Approach on Geometry," Journal of Physics: Conference Series, vol. 895 (2017)

8. Castronova, J.A.: Discovery learning for the 21st century: what is it and how does it compare to traditional learning in effectiveness in the 21st century? Action Research Exchange **1**(1), 1–12 (2002)

9. H. T. Nguyen, A. V. T. Tran, T. A. T. Nguyen, L. T. Vo, and P. V. Tran, "Multivariate Cube for Representing Multivariable Data in Visual Analytics," in Context-Aware Systems and Applications, Thu Dau Mot, Viet Nam, 2016, vol. LNICST, pp. 91–100: Springer (2016)

10. Nguyen, H.T., Tran, A.V.T., Nguyen, T.A.T., Vo, L.T., Tran, P.V.: Multivariate cube integrated retinal variable to visually represent multivariable data. EAI Endorsed Transactions on Context-aware Systems and Applications **4**(12), 1–8 (2017)

11. M. A. Weegar and D. Pacis, "A Comparison of Two Theories of Learning -- Behaviorism and Constructivism as applied to Face-to-Face and Online Learning " in E-leader conference, Manila (2012)

12. R. E. Simamora, S. Saragih, and Hasratuddin, "Improving Students' Mathematical Problem Solving Ability and Self-Efficacy through Guided Discovery Learning in Local Culture Context," International Electronic Journal of Mathematics Education, vol. 14, no. 1, pp. 61–72 (2019)

13. K. Yuliani and S. Saragih, "The Development of Learning Devices Based Guided Discovery Model to Improve Understanding Concept and Critical Thinking Mathematically Ability of Students at Islamic Junior High School of Medan " Journal of Education and Practice, vol. 6, no. 24, pp. 116–128 (2015)

14. Yuliana, Tasari, and S. Wijayanti, "The Effectiveness of Guided Discovery Learning to Teach Integral Calculus for the Mathematics Students of Mathematics Education Widya Dharma University," Journal of Mathematics Education, vol. 6, no. 1, pp. 1–10 (2017)

15. Dupri, N. Nazirun, and O. Candra, "Creative thinking learning of physical education: Can be enhanced using discovery learning model?," Journal Sport Area, vol. 6, no. 1 (2021)

16. M. S. Rahman, Basic Graph Theory (Undergraduate Topics in Computer Science). Springer 2017, p. 173

17. R. J. Wilson, Introduction to Graph Theory, Fourth edition ed. Longman (1998)

18. W. S. Cleveland and R. McGill, "Graphical Perception: Theory, Experimentation, and Application to the Development of Graphical Methods " Journal of the American Statistical Association, vol. 79, no. 387, pp. 531–554, Sep. (1984)

19. T. X. Le, "An approach to evaluating the efficiency of a solution for visualization problem," Concurrency and Computation: Practice and Experience, vol. e6322 (2021)

20. Tran, P.V., Le, T.X.: Approaching human vision perception to designing visual graph in data visualization. Concurrency Computation: Practice Experience **33**(2), 1–17 (2020)

21. Aziz, R.A., Tarmedi, E., Kusmarni, Y.: Developing Students 'Information Literature Skill Through The Application of Learning Discovery Learning Model in Social Studies Learning. International Journal Pedagogy of Social Studies **3**(1), 9–20 (2018)

22. R. D. Anggraini, A. Murni, and Sakur, "Differences in students' learning outcomes between discovery learning and conventional learning models," Journal of Physics: Conference Series, vol. 1088, no. 1, p. 012070 (2018)

23. Y. A. Warlinda, Yerimadesi, Hardeli, and Andromeda, "Implementation of Guided Discovery Learning Model with SETS Approach Assisted by E-Modul Chemistry on Scientific Literacy of Students," Jurnal Penelitian Pendidikan IPA, vol. 8, no. 2 (2022)

24. P. V. Tran, H. T. Nguyen, and T. V. Tran, "Approaching Multi-dimensional Cube for Visualization-based Epidemic Warning System - Dengue Fever," presented at the 8th International Conference on Ubiquitous Information Management and Communication, ACM IMCOM 2014, Siem Reap, Cambodia, January 9–11 (2014)
25. H. T. Nguyen, T. M. T. Pham, T. A. T. Nguyen, A. V. T. Tran, P. V. Tran, and D. V. Pham, "Two-Stage Approach to Classifying Multidimensional Cubes for Visualization of Multivariate Data," in 7th EAI International Conference, ICCASA 2018 and 4th EAI International Conference, ICTCC 2018, Viet Tri, Vietnam, 2019, vol. LNICST 266, pp. 70–80: Springer (2019)
26. Huynh, C.N.T., Nguyen, H.V., Tran, P.V., Ngo, D.N.T., Tran, T.V., Nguyen, H.T.: "An approach to selecting students taking provincial and national excellent student exams", lecture notes of the institute for computer sciences, social informatics and telecommunications. Engineering **475**, 156–161 (2022)

Author Index

© ICST Institute for Computer Sciences, Social Informatics and Telecommunications Engineering 2024
Published by Springer Nature Switzerland AG 2024. All Rights Reserved
P. Cong Vinh and N. Thanh Tung (Eds.): ICCASA 2023, LNICST 579, p. 197, 2024.
https://doi.org/10.1007/978-3-031-58878-5

GPSR Compliance

The European Union's (EU) General Product Safety Regulation (GPSR) is a set of rules that requires consumer products to be safe and our obligations to ensure this.

If you have any concerns about our products, you can contact us on ProductSafety@springernature.com

In case Publisher is established outside the EU, the EU authorized representative is:

Springer Nature Customer Service Center GmbH
Europaplatz 3
69115 Heidelberg, Germany

The manufacturer's authorised representative in the EU is Springer
Nature Customer Service Centre GmbH, Europaplatz 3, 69115 Heidelberg,
Germany. If you have any concerns regarding our products, please
contact ProductSafety@springernature.com

Printed and bound by CPI Group (UK) Ltd, Croydon, CR0 4YY
29/04/2026
02099532-0001